Karatedo Ichiro

空手道一路

Karatedo: My Lifelong Journey

FUNAKOSHI GICHIN

ISBN: 978-1-950959-88-4

Translator's Introduction

This the first complete translation of Funakoshi Gichin's book. It includes several aspects that have been previously left out.

- All References to World War Two
- The chapter order has been kept the same
- All chapter names included
- Direct translation of chapter names
- Dialoged kept as dialogue
- Significant Japanese phrases highlighted
- Followed the author's pacing with paragraphs
- Final closing remarks by Funakoshi Gichin

Karatedo Ichiro

空手道一路

Karatedo: My Lifelong Journey

FUNAKOSHI GICHIN

1956

富名腰義珍　日大慶大早大松
坂屋等の唐手師範首里出身

Early Showa Era Picture Funakoshi Gichin

Born in Shuri.

Head Karate Instructor at Japan University, Keio University, Waseda and at the Matsuzakaya Department Store. Seven members of the Keio University Karate club held a Demonstration on the roof of the Ginza branch of Matsuzakaya on March 21st 1934.

Preface

In the blink of an eye more than forty years have passed since I set about introducing the core concepts of Karatedo.

Thinking back on my life, I recall it was quite a difficult road and not at all what I had expected. Though there were times when I thought my spirit had broken, I was able to rally and end up where I am today. Nowadays, Karatedo in the global spotlight. This is due to my revered teachers, the Senpai who came before me and the support of my friends. At the same time, to those who started after me and are continuing to train in this organization, your fiery passion for refining this art is rewarding, and I am deeply indebted. I am nothing more than a single person seeking to give his perspective on the path of Karatedo. One might say circumstances favored me.

It is not an exaggeration to say that my entire life, nearly ninety years, have been spent following the path of Karatedo. I was not only born a frail child, but I was also timid by nature. Thus, from a young age, I was encouraged to join a Karatedo Dojo. From that point on, my body gradually became stronger. However, it was not until the final years of elementary school that I become hooked on this martial art. Further, I recall that my formerly introverted nature soon began to shift to a brighter, more outgoing personality.

Looking back on the ninety-odd years of my life, I discovered that all the months and years I have lived, starting when I was a youth, then as a young man and finally as an elderly man (I don't really like to refer to myself as elderly, however it can't be helped)[1] were spent following the path of Karatedo. It is, in fact, thanks to Karatedo that I have never once had to trouble a doctor with my problems. I have never taken any foreign medicines, whether powder or liquid (Western medicine,) and certainly have never had a shot. Amongst my friends I am known as having *Fushinmi*, an indestructible body. However, they are incorrect, I have *Fubyomi*, a body impervious to disease, not an indestructible body. A body that has been forged through Karate training does not leave a single gap for a malicious disease bearing demon to enter.

In my opinion there are only three types of disease that affect man. The first is the Disease of Fever, the second is Disease of the Stomach and the third is External Injury. I believe the cause of Disease of the

---

[1] Brackets are by Funakoshi Sensei.

Stomach is malnutrition and poor blood circulation. Therefore, my solution for Disease of Fever is to sweat it out by doing Karate training, and in order to stabilize your blood flow for Diseases of the Stomach, you should also do Karate. Only responding to external injuries should be handled differently. Basically, preventing external injuries requires that you remain vigilant in order to avoid dangerous situations. Those training Karate need to understand that you can't neglect your health in favor of only focusing on practicing striking with your fists or kicking with your feet.

Karatedo has only just begun to sprout, and we must diligently tend it so that it may continue to grow. Recently, not only have young men and women begun the study of this martial art but clubs for people advanced in years as well as the very young have also been started. In addition, it should be a source of great pride for everyone in the organization that Karate has been spreading overseas for some time now.

To the reader, though desiring this brings me shame, I hope that you enjoy the nonsense written by this elderly fellow. In my remaining years I hope to spur these old bones on in order to lead others to success. Through Karatedo I hope a path of great promise can be opened up for the future generations of Japan. I humbly ask for your support.

September Showa 31 (1956)
Shoto Funakoshi Gichin

Husband and wife from the Ueekata 親方 aristocracy
19th century

Translator's Note: Ryukyu Topknots

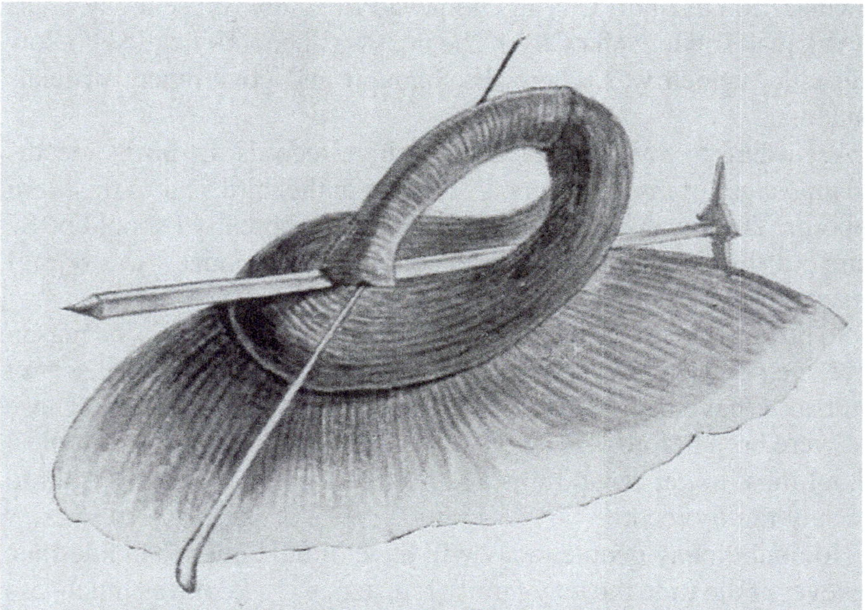

Fig. 1. Head-dress of the Men of Okinawa.

Life in the Luchu Islands
William Henry Furness 1899

In Ryukyu the Chonmage topknot was called *Kata kashira* 欹髻 Young men would get this hairstyle at their coming of age ceremony when they reached fifteen. They would also receive a hairpin indicating their status.

Stylistically, the smaller the final coil of hair the higher the person's status. Ryukyu men would often shave not the forehead, but the very top of their head (something done by some Sumo wrestlers today) in order to allow the topknot to appear smaller.

The Enlightened Party and the Stubborn Party

When speaking of the Meiji Restoration, more than ninety years have passed since the Emperor was restored to his rightful place (it is now the 31st year of Showa, 1956.)[2] I was born in the Yamakawa Town section of Shuri City in Okinawa Prefecture (present day Shuri Yamakawa Town, Naha City.) The area was generally called "below the castle" which was where the Samurai and government officials resided.

According to my family register, which records the births, deaths and marriages of my ancestors, I was born in the third year of the Meiji emperor. This means I was born in 1870, so technically I should be 87 years old (in 1956) however if you count the years since I was born, I am actually 89 years old.

So how is it that there are almost three years difference? The reason is, at one point there was a rule that to be admitted to medical school you had to have been born *after* 1870, the third year of Meiji. Those that were born before 1870 were not eligible for (to enroll in) medical school thus, to get around this age restriction, I had no choice but to alter my family record.

No doubt many people today will have trouble understanding this, however at the time the way family registers were kept was much less formal than it is today.

In the end, I was allowed to enroll (in medical school) however I encountered an unexpected obstruction which meant giving pursuing a medical degree with tears in my eyes.

At the time, only a few years had passed since the Meiji government had formed and even in Tokyo, his majesty's seat of power, men with topknots were hardly uncommon. As for Okinawa, virtually all men still wore top knots.

However, one of the conditions for enrolling in medical school was cutting off my top knot.[3] In the end, as I considered wearing a

---

[2] Brackets are by Funakoshi Sensei.

[3] In Meiji 4 (1871) a *Sanpatsu Datto Rei* Order Removing Topknots was issued by the government. This meant that government officials had to allow their hair "to grow free" when meeting representatives of other nations. In addition, they were not allowed to wear a sword and dignitaries from other nations would be received in a room prepared

traditional topknot extremely important, I ended my dream of studying medicine. Thinking back on it, I cannot but shake my head with disbelief at my decision, but in Okinawa at that time, doing such a thing (cutting off your topknot) would have caused serious ramifications.

This wasn't happening just in Okinawa, there were similar little battles occurring all over Japan as the push to civilize the nation was beginning.[4]

That was an extremely interesting era when the new political party locked horns with the old party. Each side had its own arguments which they pitted against the other party. Those two opposing sides constantly glaring at each other was the reality in Okinawa, though it seems the same pattern was repeated all over Japan.

The two sides in Okinawa were the *Kaikato*, the Enlightenment Party (progressives) and the *Gankoto*, the Stubborn Party (conservatives) and they were diametrically opposed to each other.[5]

---

with, "chairs with high legs" and a "table." Everyone would also receive guests while wearing shoes.

[4] The Meiji government's policies were aimed at modernizing the country to a Western standard. Some of the terms used were:

脱亜入欧 ***Datsua Nyuo*** Move away from Asia, Accept Europe

文明開化 ***Bunmei Kaikaka***

Civilization and enlightenment. This was a key phrase in Japan's Westernization movement.

見苦敷風習 ***Migurushiki Fushu***

Describing Early Meiji Era traditions as unsightly and shameful.

卑敷風俗 ***Hishiki Fushu***

This phrase also laments Japan's indecent cultural traditions.

御国体 ***Gokokutai***

This phrase refers to the Nation of Japan as a Whole, its image. This term was used to indicate the Migurushi-Fushu were adversely affecting the image of Japan on the world stage.

[5] The Stubborn Party was in favor of maintaining ties to China, while the Enlightenment Party was in favor of completely unifying with the Meiji Government and its goal of Westernizing the nation. Even as late as The First Sino-Japanese War (1894 ~ 1895) the Stubborn Party held an event praying for Chinese victory while the Enlightenment Party pray held and event praying for Japanese victory.

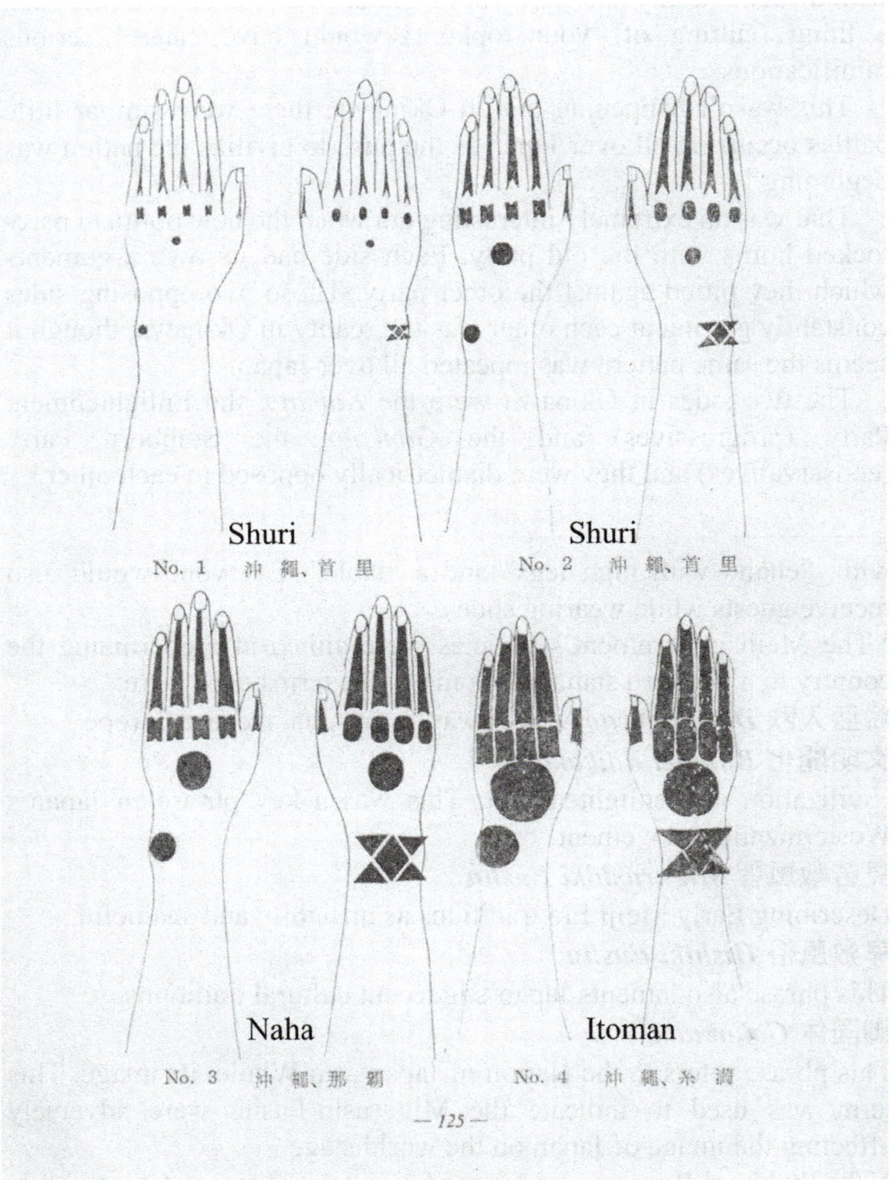

Women's tattoos from various regions
Tattooing in Okinawa Volume II
Obara Kazuo 1931

Apparently the first issue that divided the two parties was whether men should cut off their top knots or keep them. The Stubborn Party had many members from Samurai families as they refused to remove their topknots. On the other hand, the Enlightenment Party, which advocated banning the topknot, was comprised mostly of commoners with a few progressives from the Samurai class. As for myself, I wasn't really for either party, but as most of my family and relatives where members of the Stubborn Party, therefore they would not permit me to cut my topknot off.[6]

It goes without saying, but as the government had already passed a law ordering all men from the former Samurai class to cut off their topknots and it was impossible for that state of affairs to continue forever. As for me, I had worn a topknot from my coming-of-age ceremony until I turned twenty and this perhaps indicates the degree to which the two sides were opposed and how much support there was for maintaining the old-fashioned traditions and customs. When I say "twenty years" this of course means it was 1887, the twentieth year of Meiji. It's hard to put into words how much my family was against the idea of me cutting off my topknot, as they, and my other relatives, were all members of the Stubborn Party that vigorously opposed the Enlightenment Party.

Before the Meiji Restoration, my family were lower-class Samurai who served the now former head of the domain.[7] I am the only child of Gisu, my father. I was born premature what we call "a few months short" and thus I was a weak baby. My parents and grandparents said things like, "I don't think this child is going to live for long."

---

[6] The political parties also were on opposite sides of the tattoo issue. In Ryukyu all women had their hands tattooed. The tattoos differed considerably from island to island and on the main island of Okinawa, there were class difference. See the illustration by Obara Kazuo, on the previous page, from his research in the 1930s.

[7] Ryukyu was a Domain from 1872~1879, still ostensibly ruled by the King, who became the "head of the domain." The Ryukyu Domain was made a prefecture in 1879, and the name was changed to Okinawa.

弓長七尺餘卓地高齊屋簷箭比中國箭乃較短一握許射必卓地執靶時不在正中乃就下窄處扣絃發矢皆用決拾如古制　發不甚遠舊錄云射一百步外則未之見

Instead of saying, *Prevention is better than a cure* my parent's philosophy was *First is prevention, second is also prevention.* So, they basically raised me as if I would shatter into pieces at any moment. Thus, it probably comes as no surprise to learn that both my grandparents on my father's side as well as my grandparents on my mother's side fawned over me as they thought I was quite hapless. Early on I was sent to my mother's parents' house in Oyatomari.

Eventually, I began to attend elementary school. My grandfather taught me many things and also had me read aloud *The Four Books and Five Classics of Confucianism* even though I couldn't understand them.[8]

It was right after I entered elementary school that I became friends with the honored eldest son of Asato Anko Sensei, the man who would teach me innumerable things and set me on the path of Karate that I would follow for the rest of my life. I can't recall exactly when, but at some point Asato Sensei begin teaching me *Te-hodoki*, how to free a hand that has been seized.

At that time in Okinawa, Asato Sensei's technique was unparalleled, and he was renowned for being head and shoulders above everyone else in terms of ability. It goes without saying that he was also adept at the equestrian arts, sword fighting, archery[9] and other martial arts in addition to being a scholar with a profound understanding of the inner mysteries of classical literature. In short, he was the quintessential *Bunbu-Ryodo* a man who has mastered both the literary and the martial arts.

In Ryukyu during the Edo era, there were two classes of Samurai which were known in the Ryukyu language as *Udon* Great Lord and *Tonchi* Inner Lord.

---

[8] Funakoshi is referring to a learning style called *Sodoku* 素読 reading a text aloud without trying to understand it. The purpose was to develop recitation ability.

[9]Illustration of archer on previous page from *The Record of Central Mountain Kingdom* 中山伝信録（1721）by Jo Hoko 徐葆光 who visited Ryukyu as a vice-envoy in the tributary mission. In his description he states that the Ryukyuan man is shooting a bow that is "a little over seven Chi 224 cm / 7.4 feet long.

The illustration of a typical Ryukyu war horse on the following page is from the same book.

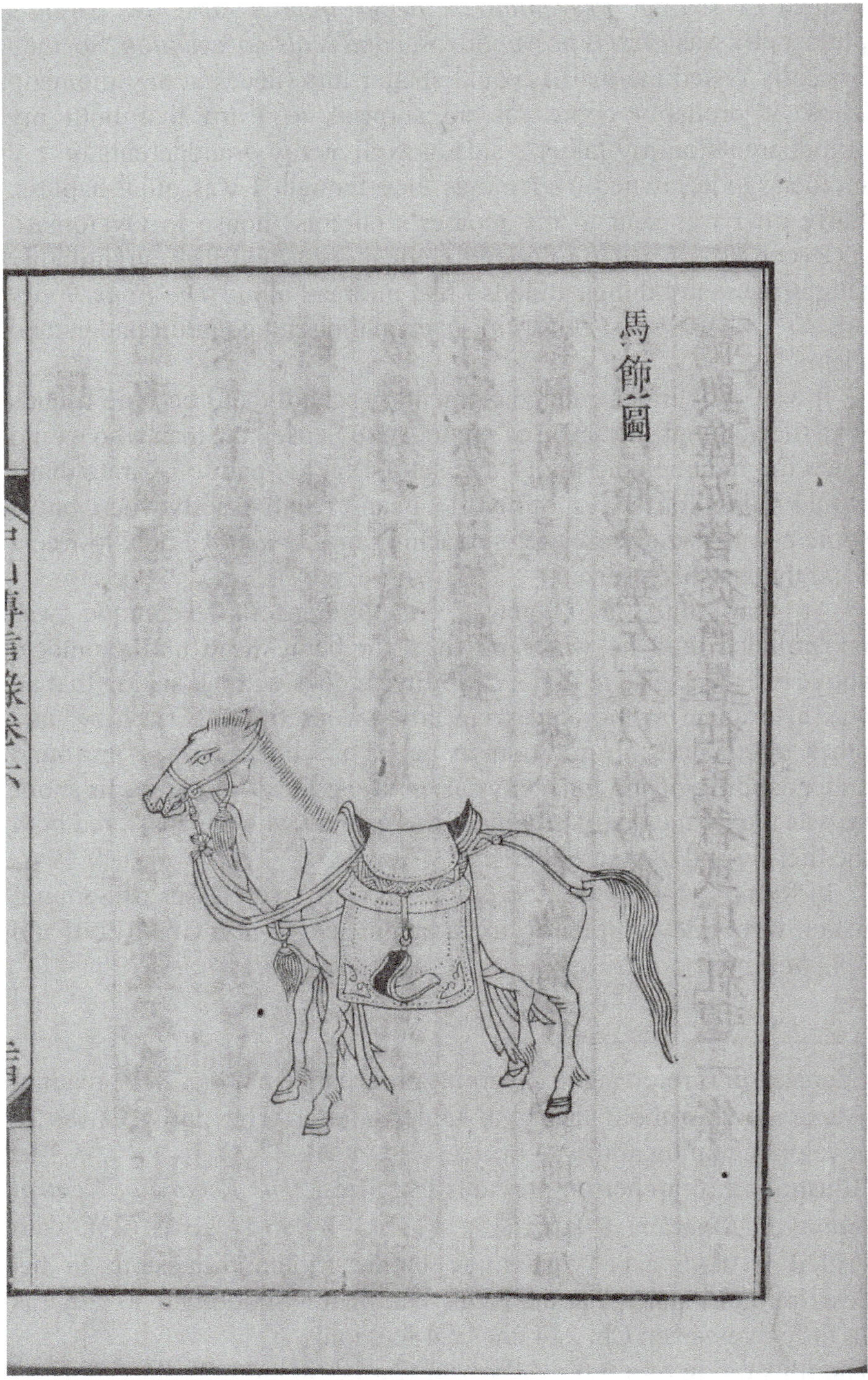
馬飾圖

A Great Lord was equivalent to a Daimyo, a high-ranking Samurai who ruled over one of the three-hundred domains in mainland Japan. As for an Inner Lord, they were equivalent to a *Chumyo*, a Samurai who oversaw a village or small town on mainland Japan. Asato Sensei served as an official for the latter, the Inner Lord who oversaw Asato Village, which was located between the cities of Naha and Shuri.

Asato Sensei was held in high regards as evidenced by the fact that when he visited the now former head of Ryukyu Domain he was received as a guest rather than as a retainer. As it turns out, I was fortunate enough to find in him a great teacher willing to instruct me in Karate. It is important to understand that at the time people were not allowed to openly practice Karate. Saying that we trained in secret sounds a bit strange, however those of us who practiced Karate at that time were very reluctant to discuss that fact with others. This will all become clear when you read this section regarding the origins of Karate, however for now understand that, unlike today, Karate was practiced in secret, hidden by the darkness of night.

Asato Sensei's house was a little over a mile from where I lived. Interestingly, even though I was still a child I did not find it troublesome to walk that distance. In fact, it became my habit every night to stroll down the road to Sensei's house and back while swinging a paper lantern light in my hand. After about two or three years of attending lessons, I began to have confidence in my formerly frail body. I discovered that Karate was extremely interesting and, as I previously mentioned, I was overjoyed to see the physical changes to my body that this martial art brought about. I had grown stronger and healthier. I can't say exactly when it happened, but at some point, I became single-minded in my devotion to pursuing this art and betting my future on it.

As I previously mentioned, I gave up my dream of attending medical school as I did not want to cut off my topknot. However, now I had to decide what sort of job to seek out. Since from a young age, when lived with my mother's parents in Oyatomari, I had been taught Chinese literature by both my grandfather and by Asato Sensei[10] I decided to try use that background to become a teacher. So, I took a test that was known at the time as the "elementary school assistant teacher test."

---

[10] This refers to studying Chinese literature in Chinese.

學生の唐手練習

Students learning Karate (Chinese Hand)
Early Showa Era Photograph

I passed and the next thirty years of my life was spent behind the teacher's desk. The first time I stood in front of a class of students was the twenty-first year of Meiji (1888) and I, according to my family register, was twenty-one years old.

However, as it turned out times were changing and to become a teacher, I would have to cut off my topknot. Though I could sense the new mood, my parents and relatives remain relentlessly opposed to Samurai giving up wearing a topknot. Unfortunately, a person seeking to teach in the new educational system would hardly be allowed to present himself with an old-fashioned hairstyle, thus one condition of becoming a teacher was cutting off my top knot. So, after being hired as a teacher I returned home dressed in the uniform of a prewar educator and announced my decision to my parents. I had the uniform jacket with a stiff raised collar, gold buttons with embossed cherry blossom pattern and a hat with the school insignia…and no topknot.

Seeing me, my father was furious and shouted, "You are a member of the Samurai class, why are you dressed like that!?" My mother couldn't even look me in the face and simply walked out the back door and went to her parents' house. Young people today probably can't imagine a scene like this.

So, during all this time my Karate training continued. It goes without saying that the training was only done at night. I don't believe I ever missed a day of training despite having a full-time job as a teacher. This naturally led to questions from my coworkers, "Mr. Funakoshi you sometimes don't get home till morning!"

There was also gossip about me too, "I'm not sure why, but Mr. Funakoshi will get home from school, but as soon as night falls, he leaves again."

At the time training consisted of repeating one Kata, not for several months but for several years. We would have one or even two years of repeated practice before Sensei would say *Yosh!* Good![11]

---

[11] *At the time training was completely unlike the training done today, we would work on one Kata for three years. In other words, to go from learning Heian Shodan to Heian Sandan would take a decade of training. Nowadays instruction in Karate is still strict, however it doesn't compare to how it was done back in the day.*

-Gima Shinkin
Monthly Karate
May 1957

The purpose of this type of training was not to memorize the technique but to repeat it until we discovered the inner mystery hidden within that technique and make it part of ourselves. Only then would Sensei say *Yosh!* Good!

The severity of training is something that is hard to put into words. It was as the sayings go,

Dojo no Yukaita wo Nameru
Licking the floor of the Dojo to learn the important lessons

And,

Yukaita wo Seotta
Carrying the floorboards on my back

Thus, in the garden of each Sensei, which served as their Dojo, I did as the phrase suggests and licked the dirt to absorb the difficult and important lessons and loaded them onto my back.

When training at Asato Sensei's house, I would be in the middle of the garden and he would watch me from the edge of the open porch. Despite being advanced in years, he would watch me intently while dressed in a Hakama and sitting in *Seiza,* without so much as a cushion.

Whenever I finished a Kata all he would say is, *Yoroshi*, Good Enough, or *Yosh*, Good, or *Mo-ichido Yare*, Do it again. I never expected him to grant me a compliment; I simply repeated the techniques repeatedly until I was soaked in sweat.

Asato Sensei always kept a lantern beside him as he sat in *Seiza* watching me. I would be so exhausted by the end of training that the feeble light from his lantern would barely register in my eyes. The training would often extend until nearly dawn. However, after training was done, he would chat with me like an affectionate father and helped me to understand what the spirit of Karate was all about.

I still vividly recall something Sensei always said,

Ima Shibaraku
We will continue this for the foreseeable future

What he meant was that we would be continuing to study whatever Kata we were training at the time, thus we never asked Sensei if he would teach us a new Kata or made other such requests. Around this time, in addition to receiving instruction from Asato Sensei I also began learning under the tutelage of Itosu (Anko) Sensei, a dear friend Asato Sensei.

Itosu Sensei was also a man who was a frequent visitor to the household of the now former head of Ryukyu Domain. He was also known equal in ability to Asato Sensei.

I had the privilege to sit with them once and was able to ask them both questions about Karate. Suddenly they asked me, laughing, "Young Funakoshi, exactly who's student are you?"

Truth be told, I would have been unable to part with either Sensei. If either of them had disappeared when I was training, I think my Karate would have ended up incomplete and ineffective.

Though it has been more than eighty years since I embarked on this path, I still feel like a beginner, a path that I cannot imagine would have been possible if I hadn't been blessed by the teachings of these two Sensei.

Being a Good Talker Doesn't Make You an Expert

So then, for those of you not familiar with Karate, I would like to say a few words about, "What Karate is not." However, while I would like to take the time to carefully explain what Karate is, I must first address some big misconceptions regarding this art. Thus, my initial comments will be about, "What Karate is not."

When describing the powerful strikes in Karate, there are some practitioners that explain them like this:

"There is a Kata called *Nukite*, Piercing Hand, that involves slamming all five fingers into your opponent's side so that the tips of your fingers pierce through the skin into the flesh. You then seize your enemy's rib and yank it out. The training you must do to achieve this strike is no easy feat. First fill a bucket with two or three *Kin*, 2.5~4 pounds, of beans.[12] Then practice slamming your hand into the bucket with your fingertips joined together."

"At first, the skin on your fingertips will split and bleed from repeating this action, thousands of times, then tens of thousands of times a day. However, as your fingers harden, your hands will begin to change shape. After even more repetition that level of training will become mundane.

---

[12] Kin 斤 is an archaic unit of weight. 1 Kin is 1.3 pounds/600 grams.

Having graduated to the next step, you then fill the bucket up with sand instead of beans. After switching from beans to sand, you will find it is harder for your hand to pierce through, however with days and months of practice, you will eventually be able to stab your fingertips down through the sand to the bottom of the bucket."

"After graduating from sand, you then switch first to gravel and then to small stones, before finally training on a bucket full of lead shot. Having completed this course, you will be able to slam your fingers through a board, break rocks and easily punch through a horse's stomach."

This would seem to be the kind of claim I shouldn't have to bother refuting, but apparently there are no shortage of people who believe it. Karate practitioners, as a whole, cannot be said to not bear some responsibility for the spread of such rumors.

When asked by a person, "You look like a real Karate practitioner. Can you smash a rock with your fist or punch holes in a person's stomach with your fingertips?"

As the questioner clearly has no idea about Karate, it is best to reply along the lines of, "Absolutely not, Karate isn't comprised of parlor tricks." However, some people see fit to answer with, "I can't say that never happens." This, to the uninitiated, makes it seem like Karate is something unobtainable and they become fearful of it.

Invariably the problem stems from braggarts who, coincidentally, seem to have the best way with words. This means that, due to the braggart's deftness with words, listeners are left convinced what he said was true. Such braggarts may very well think by explaining things convincingly, he or she is gilding the name of Karate, but in reality, they are dragging down the reputation of Karate.

It is possible there may well have been some Karate practitioners living in the past who could perform such a feat. However, as far as I know, there is no such person alive today. Also, there are people that say things like,

"Grip strength is important in Karate that's why we train using jars filled with sand. Taking one in each hand gripping the lip we train grip strength by swinging these jars around. By doing this sort of training we can yank the flesh free from their bones when we seize the hand or foot of an opponent."

While this is possible, to a degree, the implication that Karate practitioners can rip a person's flesh into a thousand pieces with just their hands is a wild exaggeration.

貫き手

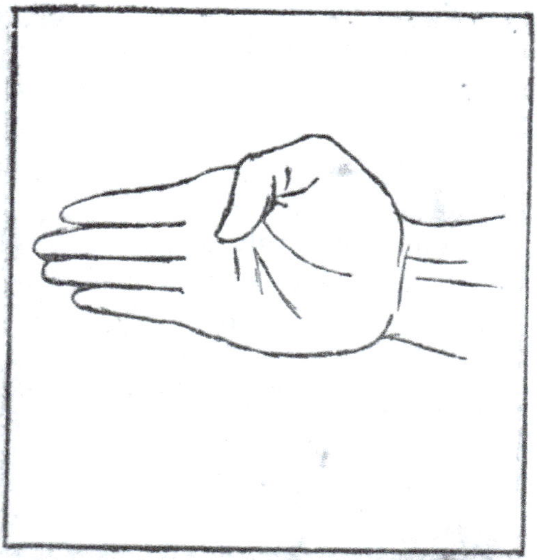

上圖の如く四本の指を眞直に伸ばし拇指を曲げる、俗にこの種の手を貫手と名づく

As the illustration shows your four fingers are extended straight out with your thumb bent. This type of strike is known colloquially as Nukite, Piercing Strike.

Karate: Ryukyu Kenpo 唐手：琉球拳法
Funakoshi Gichin
1922

Takagi Jotaro, former head of the Shotokan, mentions that Funakoshi Sensei, who spoke of the importance of tucking your thumb in, quoted him as saying,

Be careful not to allow your enemy to grab your thumb, if your thumb gets grabbed, you won't be able to do anything. That is why you tuck your thumb against your palm when making a Shuto.

Takagi Jotaro, former head of the Shotokan.
2004

I can't recall exactly when, but a man professing to know the secret technique known as *Niku Sengiri*, shredding the flesh into a thousand pieces technique, came to my Dojo saying he would teach us this method. We immediately accepted. Though he demonstrated his method, its effectiveness was rather meager. When the visitor applied the technique on me, it felt no more than a harsh twist. While the technique was supposed to rend flesh into a thousand pieces, in reality it was only a pinch that barely left a bruise.

In short, his grip strength hardly surpassed that of an average person. I will however tell you the story of a man who possessed great strength. A man I know could hold onto the rafters of a house and move hand over hand all the way around it. However, if you are not familiar with the construction of Okinawan houses then you might not know exactly what this means.

It is true that some people like my revered teacher Itosu Sensei were very strong. I saw him seize a piece of tortoise-shell bamboo and shatter it in his hands with a *Bari-bari* sound. But as to how he got so strong? If I had to guess, I would probably surmise that it was a gift from the gods rather than the product of training.[13]

It goes without saying that if a person trains their body consistently, they will reach a high level of fitness, but it is important not to forget that there are limits.

Regarding breaking boards, your average person would have trouble simply breaking one, while Karate practitioners can split a dozen thick boards. Karate practitioners can then follow up by breaking several roof tiles stacked one upon the other. That being said, with training, anyone can reach this level of proficiency. However, understand that *Ita-wari*, board breaking, and *Kawara-wari*, rooftile breaking are simply two types of *Tameshi*, tests, and do not represent the true nature of Karate. Further, they are in no way secret techniques. In fact, they could be thought of as being altogether outside the scope of Karate.

Those unfamiliar with Karate often ask silly questions like, "How many boards do you have to break to get to this Dan rank?"

However, ranking in Karate has nothing to do with how many boards you can snap in half or how many roof tiles you can shatter. Karate is an exceptional martial art so if anyone begins bragging about

---

[13] https://dl.ndl.go.jp/pid/12146816/1/133?keyword=漢那庸森

how they can break boards, roof tiles, yank out ribs or shred flesh into a thousand pieces or any other minor details not pertaining to the core ideals, you can be sure that that person knows nothing of Karate.

Practical demonstration of Karate: *Ita Wari* board breaking
Early Showa Era

Memories of My Revered Teacher

- **The Merits of Cutting off the Top Knot**

When I was first hired, elementary school teachers were divided into four categories based on license:

- Fully Licensed Elementary School Teacher
- Standard Course Licensed School Teacher
- Specialist Teacher
- Assistant Teacher

As I mentioned before I got my start by passing the test to be an Assistant Teacher. Quite soon after that I passed the test to become a Standard Course Licensed School Teacher. Having received that certification I was moved from a regional school to Naha, which was directly administered by the prefectural government. From that point on I wielded my teachers cane in front of a classroom in Naha city at the heart of the prefectural government.

This was a lucky break for me, as far as Karate training went, though on the other hand maybe this transfer was due to Karate. I was a Standard Course Licensed School Teacher. At that time to meet their requirements for compulsory education students had to attend first through fourth grade. However, I only had a license to teach first and second grade. To teach 3rd ~ 4th grade as well as 5th ~ 8th grades one needed to be a Fully Licensed Elementary School Teacher.[14]

Later I took the test enabled me to teach all grades of elementary school students, however for a time I was only able to teach up to the second grade.

Though there was a method of advancement through standardized tests, the number of teachers graduating from teacher's college began

---

[14] In 1875 the Japanese government opened 24,303 elementary schools and enrolled 1,928,152 students. In 1856 compulsory education became 1st ~ 4th grades with 5th ~ 8th grades being optional. Initially, the school was not free but by the early 1900s the fee was waived and enrollment increased. Japan was able to rapidly implement the school system as there were already tens of thousands of *Terakoya*, temple schools, already providing education to the children of townsfolk and farming families.

increasing, meaning advancement for people with my background was becoming more difficult.

That being said, I did receive an offer for a senior position from a school principal. Unfortunately, as the offer of advancement came with the requirement that I transfer to one of the outlying islands or a remote area, I refused. I did not want to be separated from my two teachers, Asato Sensei and Itosu Sensei, so I lost that chance at advancement.

That is one reason why I spent most of my career in schools in the central part of Naha. There was also another big reason.

The second reason is regarding the *Danpatsu Mondai*, Cutting off the Topknot Issue. There were still members of the Obstinate Party that refused to cut off their top knots, thus there were a great number of elementary school students wearing topknots. Despite the fact that it was well into the Meiji Era, being the 24th or 25th year (1891 ~ 1892,) the government edict banning the top knot had not fully permeated areas far from Tokyo.

The ministry of education, as well as other government offices, were not going to ignore the problem and they continued to try and convince children holding out of the necessity of cutting off their top knots.

Many of the children had long obstinately refused to cut their hair, with the result being that many of them delayed their enrollment in elementary school. This meant that when they finally began attending school many were older and larger than average. In addition, by this point Karate was being practiced more openly than before. The result was your average teacher couldn't handle the children. Scuffles between students and teachers were not uncommon. As for myself, I was born into a household of Obstinate Party members and only belatedly cut off my topknot when I was around twenty years old so I could well understand the children's resistance. That being said, we were at the peak of the Meiji Era and it was hardly the time to be debating whether or not to keep the topknot. Thus, as I had a knowledge of Karate, I became one of the teachers active in cornering the children and taking scissors to their topknots.

During that roundup, some of the children we cornered faced off against us in Kamae, eyes brimming with tears and fists clenched until we finally got the scissors on them.

As some of the children were actually quite skilled with their fists, taking it for granted that an adult could control them would lead to a tough situation. At some point this whole issue came to an end and all the boys at school had *Maru Bozu*, nicely shaved heads like little monks.

Memories of My Revered Teacher

- **Regarding Asato Sensei**

While I previously spoke briefly about Asato Sensei, would like to expand on that. He was extremely insightful regarding government policies and always seemed to know which ones would be enacted. He even served as an adviser to the former head of the domain. One time he told me,

"Young Funakoshi, once the Siberian railway is completed, Japan and Russia will go to war." At the time I had no idea why he was telling me this. Looking back on it now, I realize Asato Sensei had a discerning eye for the current state of national and international affairs created during the Meiji restoration. When the Ryukyu Domain was going to be made a prefecture, Asato Sensei advised the former head of the domain to quickly cooperate with the new government.

Further, in response to the changing times, he led by example and cut off his top knot. Conditioned to being expert at Karate, Asato Sensei was also adept at sword fighting. I believe it was the Jigen school of sword that he studied. I heard that he was quite confident in his ability, though he despised people who talked with *Tai-Gen-So-Go* speech full of boastful talk and grandiose claims.[15]

Asato Sensei was known to sometimes boast about his own prowess, albeit in a joking manner, "If anyone wants a *Shinken Shobu*, to duel with live blades, then I am ready anytime!"

The truth is that Asato Sensei was quite skilled. To understand his ability, consider the following story. Once he challenged a man named Kanna Yorin, who was very proud of his Kendo ability, to cut him with a real sword. When Kanna cut, Asato Sensei moved his arm in a flash, knocking the attack aside and planting his body firmly against his opponent's side, preventing him from moving. Kanna was also a famous Bushi, or Samurai.

In Okinawa when we use the word Bushi, we are not simply describing a person from the Samurai class, rather, it implies a person who is an expert at martial arts. Kanna can best be described as "a man

---

[15] *Tai-Gen-So-Go* 大言壮語

who through his knowledge of Japanese and Chinese literature has achieved an overreaching knowledge of the world."

He was a big man covered with great mounds of muscle. In particular the muscles around his shoulders made him look like a bull. People used to compliment his strength by singing, "Kanna's shoulders are two stories high!"

However, even though Kanna Sensei was a man of unparalleled bravery, possessed a hot passion like a ball of flame and was respected by all the great masters at the time, when he faced off against Asato Sensei neither his hands nor his feet could be put to any effect.

Asato Sensei would frequently talk about this episode:

"You can divide people into three categories when you are facing off against them:

Man (Overwhelming Your Opponent)

Sun (Attacking Carefully)

Etsu (Attacking with Deception)

Mr. Kanna was clearly the *Man* (Overwhelming Your Opponent) type as he was unparalleled at applying ferocious pressure on his opponents. His initial move was to ensure his opponent felt intimidated thus I knew that I should respond with either *Sun* (Attacking Carefully) or *Etsu* (Attacking with Deception.)

Since from the very outset of the duel with Mr. Kanna, I could sense the *I'm going to attack, I'm going to attack* intent emanating from him, all I really had to do was open a gap in my defenses in order to draw him into striking. I don't think I was constantly consciously making that decision, but at any rate he leaped in and cut. In the end, he fell perfectly into my trap."

I was fortunate to hear Asato Sensei talk about this duel and other important points, one of which he repeated frequently,

"People that train Karate should think of their hands and feet as swords. An opponent should know that if he touches them, he will get cut."

Translator's Note: Jigen School of Sword

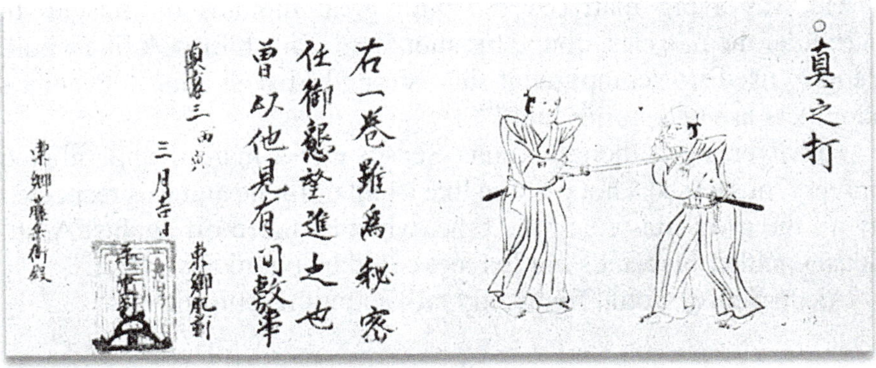

Jigen Ryu Catalogue of Transmission

Tohshin Ittai The Body and Blade as One

Jigen Ryu training differs from typical Kendo training in that you are not striking another practitioner. Instead, you are focusing all your strength into one attack, cutting as if your body and the blade have become one. In one powerful cut you are aiming to split your opponent in two. In short, this school of sword teaches there are only two possibilities in combat: either you cut your opponent down in a ferocious attack or, if you do not, you yourself will die in the attempt.

健児社と薩藩士風

The Raising of Strong Children and the Samurai of Satsuma (1943)

by Okubo Ryu 大久保竜 (1892~1943)

Translator's Note: Jigen School Techniques

Funakoshi Gichin does not define the terms Etsu, Sun and Man, however they are described in an Jigen School of Sword scroll.

東郷示現流目録初段～四段

Togo Jigen School of Sword Catalogue of Techniques 1st ~4th Rank

「越」の打法 How to Attack With *Etsu*

This attack begins from Future Dragonfly stance. The moment it seems you are going to strike, you refrain. However, you continue to advance. Without moving your right fist, pull your left elbow back and cut through either your opponent's left or right wrist. When drawing your sword back, don't draw it straight back but rather strike near the base of your opponent's right wrist as you withdraw. Understand this is not a vertical strike but a horizontal strike.

「寸」の打法 How to Attack With *Sun*

You are holding the handle of your sword with both hands. Pull your wrists against your chest. Your sword should be held vertical with the Tsuba, hand guard, should be about two inches away from your chin. Stand straight, with your feet spread evenly and facing your opponent, slightly closer together than the basic stance. Begin your attack by stepping forward with your right foot, being sure to keep your left heel on the ground. Rotate the end of your foot, while keeping the handle of your sword against your chest. Then raise your sword and cut down aiming for a spot below your opponent's sword guard, ensuring your left hand ends up against your left hip bone.

「満」の打法 How to Attack With *Man*

To attack with Man, filled to the brim, rapidly close the distance to your opponent by leaping in. When striking with your sword, keep your body straight and your right foot planted on the ground. Bring your left foot up beside your right foot before leaping forward with your right foot and striking.

After reading the descriptions of these techniques, I have defined these styles of fighting as follows:

Man	満	"Filled to the Brim"	Overwhelming Your Opponent
Sun	寸	"Slightly"	Attacking Carefully
Etsu	越	"Overcome"	Attacking with Deception

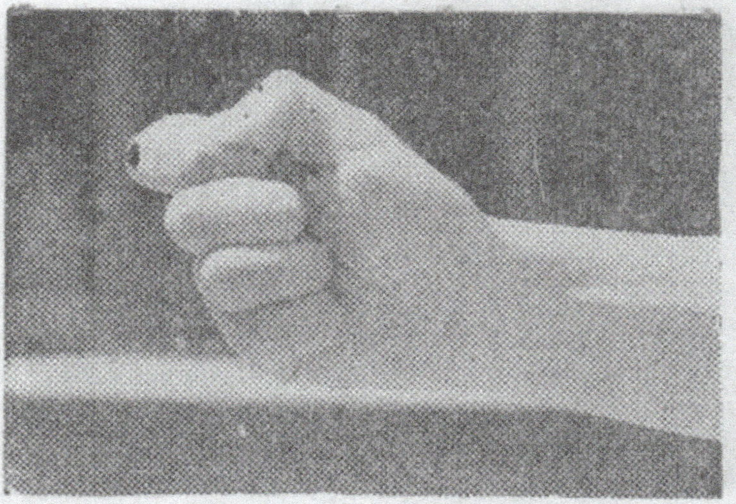

人示一本拳

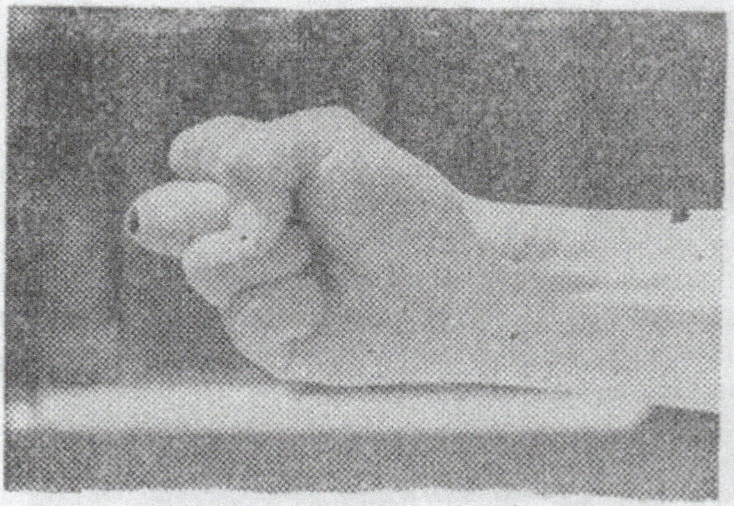

中高一本拳

Top: Hitosashi Ipponken
Bottom: Nakadaka Ipponken

Karate no Naraikata 空手の習い方
By Oya Reikichi 大家礼吉
1955

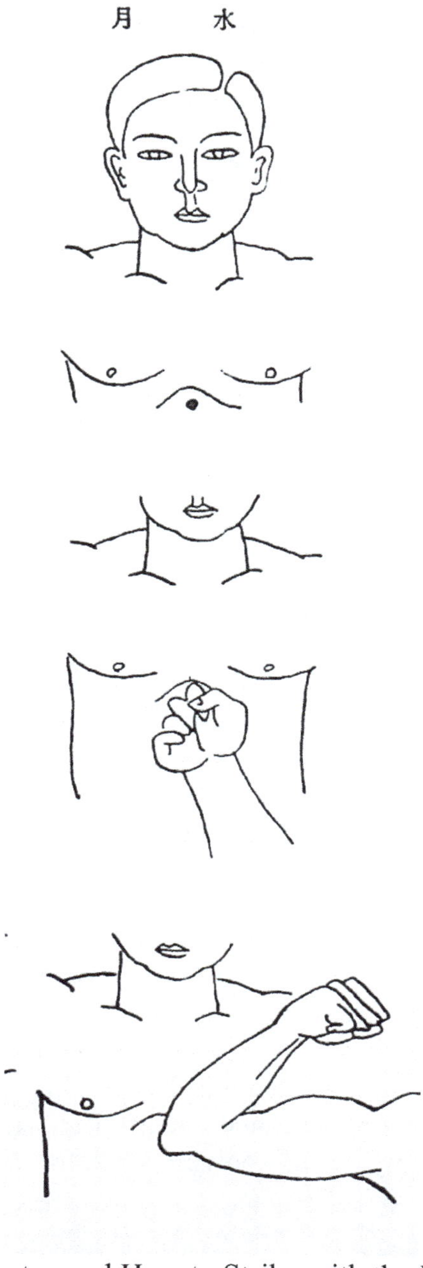

Location of Suigetsu and How to Strike with the Fist or Elbow
Kenpo: Illustrated Explanation:
Secrets of Striking, Killing, Resuscitating
拳法極意當身殺法活法明解
Fujita Seiko 1958

Having seen Sensei's Karate firsthand, I can fully appreciate what is meant by the phrase hands and feet like swords.

One day a student asked, "What is the best way to use *Ippon Ken*?" Upon hearing this question, Sensei got lightly to his feet and said, "You, try and hit me."

The questioner threw a full-force punch straight at Asato Sensei, who responded by smoothly avoided it. He then struck with *Ippon Ken*, driving his punch forward, over his attacker's forearm and forcing it down before continuing forward. His fist stopped just before Suigetsu, the solar plexus,(also called *Mizu Ochi,* water drop.) The speed at which Asato Sensei executed the technique could only be described as being nearly divine in nature, further his opponent did not have the slightest opportunity to draw away.

Asato stopped his strike just short of his target in what is known as *Kami Hitoe*, a distance so small only a sheet of paper separates them. If he had not, the attacker would surely have drawn his last breath.

Later, when the man who had asked the question was changing, we saw he had a huge purple bruise on his forearm where Sensei's strike had scraped across it and forced it down.

Asato Sensei had an almost encyclopedic knowledge of Karate masters in Okinawa. Not only did he know their first and last names, but also where they were from, who their teacher was, their level of strength, the technique they were most adept at and, quite surprisingly, the technique they were not good at. He explained to me why he had gathered and memorized all this information.

"No matter when or where you may be or what technique you are being attacked with, you must immediately take note of who is around you. Also, what kind of people are they? How strong do they appear? What is their best technique? Their worst? If you gathered this information, you will not be frightened in the least. If there is ever a person that catches your attention it is essential to make a point of getting to know more about them one of the secrets of martial arts is knowing your enemy means knowing yourself." [16]

---

[16] This seems to be a paraphrase of the famous line from *Art of War* "Know your enemy, know thyself, and you shall not fear a hundred battles"

Memories of My Revered Teacher

- **Regarding Itosu Sensei**

While Itosu Sensei and Asato Sensei were good friends, it is amusing that they both share the same first name, Anko.

As for their appearance, on one hand while Asato Sensei was tall and broad shouldered with intense flashing eyes, recalling the Samurai of old, Itosu Sensei, on the other hand, was of average height with a thick chest so he seemed like a barrel of a man. Though Itosu Sensei had a long beard, he was so mild and friendly in appearance that he almost seemed like a child. As for his strength, at that time in Okinawa there was no man who could equal him. At drinking parties, Itosu Sensei would make Asato Sensei arm wrestle him. Almost immediately after the competition began, Asato Sensei would break into laughter saying, "It's no good, unless I can use both hands, I'm not going to be able to win!"

Okinawa and arm wrestling is different from how it is done in Tokyo. Both combatants keep their fists clenched and crossed their wrists together and then begin pushing inward. In other words, this is training for an inside block.[17]

---

[17] Arm wrestling in Okinawa is called *Udei Ka Kiin* in the Ryukyu language and can be written with the Kanji 腕相撲 which is read *Udezumo* on mainland Japan. It can also be written as 腕掛 which is read *Udegake* on mainland Japan. This is a game usually done by Bushi, referring to Karate practitioners, so it is not so common amongst the populace. Combatants face each other and make a fist and join the inside of one wrist to the other person's wrist with elbows bent. They then push and whichever succeeds in pushing the other over, wins. When doing this the elbows are not resting on Tatami mats, table or otherwise supported.

There is also another version called Nobigake, extended arms, where you do arm wrestling with the inside of the wrists, but the arms are extended straight out and the elbows are not bent.

-Complete Guide to Okinawan Traditions
沖縄風土記全集
1969

I mentioned before how strong Itosu Sensei's grip strength was, next I would like to introduce the story about that. One day Itosu Sensei had gone out to the Tsuji area of Naha. Tsuji was a place where you could go to enjoy the "three businesses" which are restaurants, drinking establishments and places to hold banquets. Sensei was about to pass casually into the door of a particular restaurant when a young ruffian who had been lurking to the side suddenly shouted a Kiai of *Ei!* and launched a well-aimed punch at Itosu Sensei's side.

The attacker perhaps expected his target to twist his body away… however that is not what happened. Without even turning around Itosu Sensei responded by focusing all his power in Tanden (a spot about 1 Sun, 1 inch below the navel,) and the attacker's fist simply rebounded off him.

The young ruffian exclaimed,

南無三

Namusan! I failed!

He then started to take a step back. But just then he felt a vice like grip on his right wrist and his arm was yanked up. His attack having failed, the young ruffian's forehead broke out in sweat and he was suddenly rendered unable to breathe. Itosu Sensei, his face completely calm, twisted the young ruffian's arm up behind his back and marched him through the front door of the restaurant to a table at the back. Seating himself at a table Itosu Sensei ordered sake and snacks from the bewildered-looking serving girls.

With the ruffian's arm still in a lock, Itosu Sensei picked up a cup of sake with his left hand and took a sip before slowly twisting the man's right arm causing the man to stagger around and face Itosu Sensei. Finally coming face to face with his attacker, Itosu Sensei smiled and laughed, commenting, "I can't imagine why you have such a grudge against me, but never mind that, let's have a drink together!"

And following these words, he placed a cup of sake before the young man. It goes without saying that the previously violent man stood gaping without being able to speak as if he had been poleaxed. The point of this story is that Itosu Sensei had continuously forged his body to the point that he was *Fushimi*, Immortal.

Your average person trying to test his strength by taking a swing at Itosu Sensei was not going to have any effect.

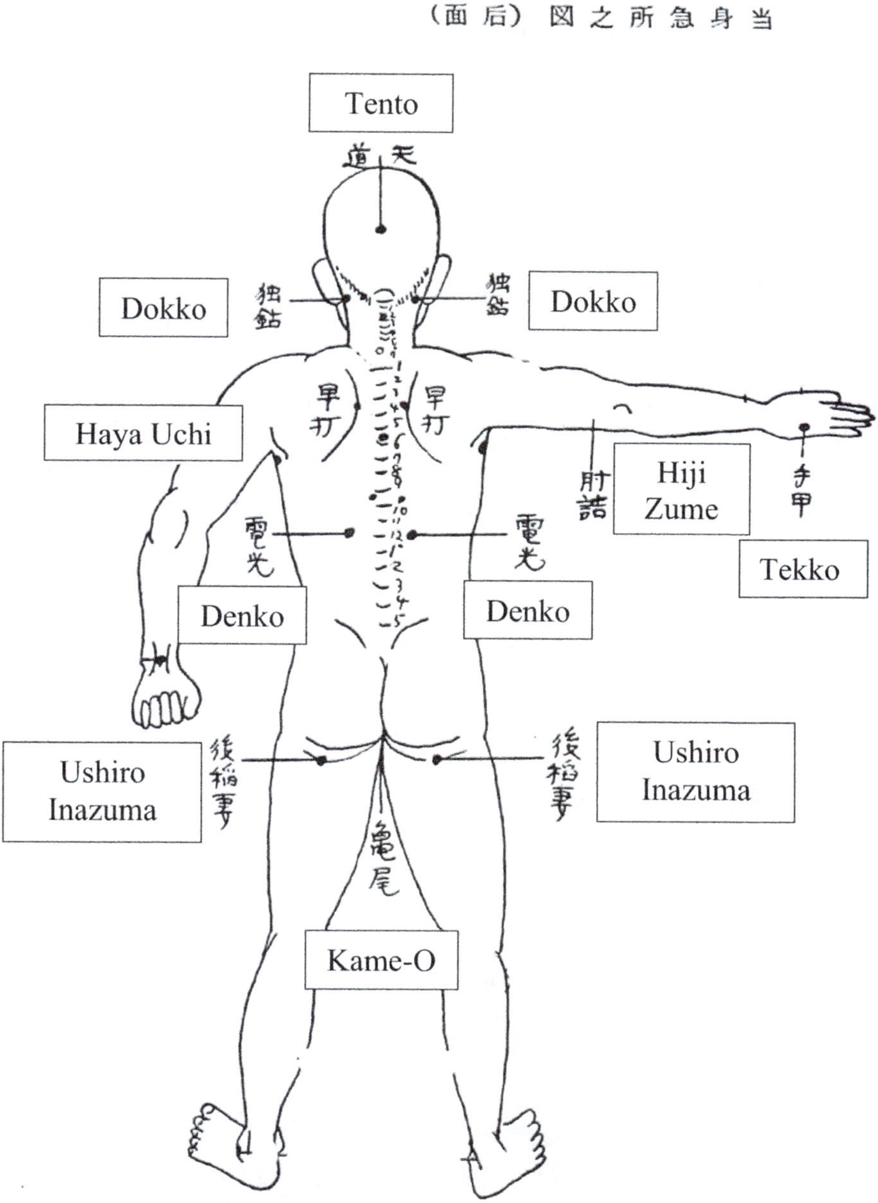

Striking Point Chart
Kenpo: Illustrated Explanation: Secrets of Striking, Killing, Resuscitating
拳法極意當身殺法活法明解
Fujita Seiko 1958

There was a certain man who served as a Karate instructor at a school I worked at. He was famous for getting into fights when he was a youth. The man liked to do something he called *Tameshi Tsuki*, a test punch. He would wait at intersections and challenge people. However, one day he found himself full of vigor from his recent successes and got the idea in his head that he would test his strength against Itosu Sensei, saying,
"I don't care how famous he is, if I strike when he isn't paying attention then I will surely win!"

So, one night, he allowed Itosu Sensei to walk past him before launching a powerful blow directed at the *Kyusho*, vital point, called *Denko*, lightning bolt.[18]

As it turned out, Asato Sensei neither adjusted his stride nor turned around, it seemed like he simply timed the speed of the punch and seized the attacker's wrist in a vice-like grip. He then yanked the man towards him and wrapped his other arm under his armpit.

---

[18] The Keio University Karate manual lists a spot called *Ushiro Denko*, Lightning Bolt on the back:

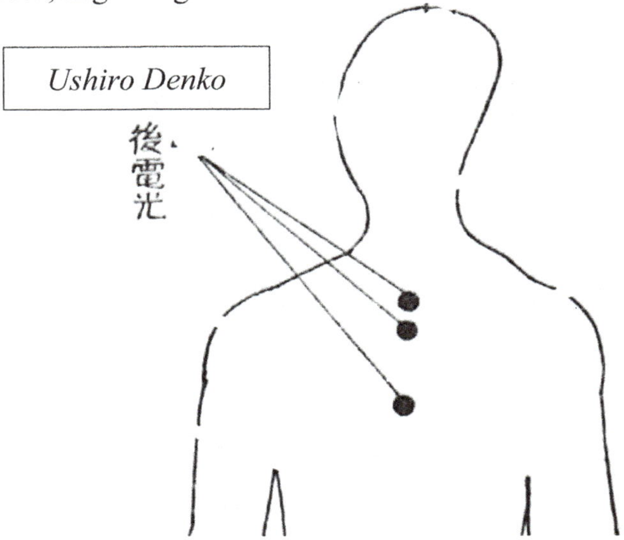

Place: Third, fifth and sixth vertebrae
Effect: Instant death
Reason: Shock to the lungs, heart, internal organs and spinal fluid.
How to attack: Ken (fist,) Enpi (elbow,) Shuto (knife hand.)

Keio Gijuku Tai-ikukai Karatebu 慶応義塾体育会空手部
1936

Consider the situation the man was in. He had been wrapped up by a man of ferocious strength who can crush tortoise-shell bamboo with his bare hands. And now that crushing grip was causing pain to flash like lightning through his torso. The attacker's right hand was numb and he could neither shove his fist forward to strike, nor pull it back to free it. With his remaining hand he desperately hit his captor, but it seemed to have no more effect than a fly. Asato Sensei then dragged the man along with him.

The man began whining "forgive me, please Sensei, forgive me!" with his arm still locked around the attacker's armpits Itosu Sensei continued casually on his way. Eventually he addressed the man in a quiet voice, "So, who exactly are you, young fellow?"

"My name is Goro..."

Hearing this Sensei turn to face the young man for the first time.

"Ah, Goro, well it is bad manners to play tricks on the elderly!"

And with a laugh he let the man go. That was how Sensei handled things.

Memories of My Revered Teacher

- **Other Memories**

While Asato Sensei advised, "Think of a person's hands and feet as swords." Itosu Sensei recommended, "Even if a person hits you, as long as it doesn't hurt who cares?" You can understand by these statements how each of these teachers perceive Karatedo.

It goes without saying that Itosu Sensei didn't mean to simply forge your body so that you could endure any punch your enemy through you. He did train his fists both night and day never ignoring his training period here is a story about him.

One day Itosu Sensei was set upon by a group of violent men. Of course, Sensei was able to deal with them easily however a witness to the fight immediately reported what had happened to Asato Sensei. Asato Sensei listened, but before the witness had even finished talking said "So all the villains ended up laid out face down on the ground. Isn't that right?"

Surprised, the witness replied, "Yes, that is what happened, he knocked them all flat on their faces, but how did you know that?"

Asato Sensei replied, "A person who has trained Karate extensively would never do something so cowardly as to strike a person from behind. If an unexperienced person were to attack him from the front,

that person would of course be knocked onto his back. However, if someone were to take a punch at Itosu Sensei, his response would surely cause the attacker to end up face down on the ground."
He then added, "It is unfortunate but there is no hope for them."

In addition, there is a tale regarding striking power that surprised Asato Sensei. This is a story that I heard directly from Asato Sensei.

One night, Itosu Sensei heard a strange noise from his front door. He quietly got out of bed and went to take a look. He realized the sound wasn't a natural one and he sensed that the noise was very much like that of a burglar fiddling with the lock from the outside.

Itosu Sensei slipped quietly up to the door and then with a shout of *Ya!* struck the wooden door with a punch like a bolt of lightning. The door was almost 3/4 of an inch thick but his fist smashed straight through it. He then used his ferocious grip to seize the burglar's wrist and then dragged the man's arm back through the shattered hole in the door. You would expect the hole in the wooden door to be cracked on the top and bottom left and right if someone punched through it however when struck by a true Karate practitioner the hole was the same size as his fist.

I was blessed to receive training and life lessons from these two Sensei, but there is one important thing I want to share with those following the path of Karatedo. I light a candle and offer incense at the family Buddhist altars for not only my two teachers but all who have instructed me. I vow never to use my fists for anything that would bring shame to a person on the path of Karatedo. This is what I do, and I recommend you do it as well.

Though I was unworthy of their instruction while my two teachers lived, they treated me like one of their relatives even after I got married and had children of my own. I believe it was due to the way I paid my respects at their family shrines that I was able to maintain these close relations.

I would frequently bring my children when I paid a visit to see one Sensei or the other to ask them a question. Whenever I humbly asked one Senesi or the other to elaborate on, or give me an opinion regarding a Kata, they would both respond in the same manner.

Standing they would tuck the hem of their Hakama under their belt, thereby allowing me to see their legs. Then turning to one of my children one or the other would say,

"Hey there young fellow, let's try one of these Kata out."

And with that, would step into the garden and demonstrate the technique using my child as a training partner.

That is how my teachers would answer my questions. I was further blessed by the fact that my children were receiving direct instruction from both my teachers. Again, I was filled with a sense of gratitude.

Both Asato Sensei and Itosu Sensei treated me the same way. They even astounded me by saying,

"You are just like us; you have a comprehensive understanding."

So, with this almost family-like connection, my children began to follow in my footsteps walking to Asato Sensei's or Itosu Sensei's house while swinging a paper lantern for light. I am incredibly grateful that that two generations, both father and sons, could receive personal training from them. This is a little embarrassing to admit, but my household was very poor. For treats all I could give my children were roasted sweet potatoes. However, both Asato Sensei and Itosu Sensei were from well-to do households, and as they were both advanced in years, they doted on my children as if they were their own grandchildren. Whenever we paid a visit, they would invariably give my children the most wondrous and unusual candies. My children were certainly overjoyed at this, but at some point, they also realized they were receiving personal instruction from the greatest Karatedo practitioners alive at the time.

One thing I would like to make special mention of regarding my two teachers is how they would mention the names of other teachers and would tell me it was imperative that I learned this point or that point directly from that teacher. They would even order me to go and have such and such Sensei tell me about a certain principle. Their attention to such details was remarkable. Typically, a teacher would be less than thrilled if one of his students was being taught by another teacher, however both Asato and Itosu would tell me to go here or tell me to go there.

While I firmly believe their humanistic approach is something that must be studied further, it is their humility that astounds me.

As far as how my teachers learned Karate, all I know is what I heard from other people. While there are many famous tales of their exploits, many of which sound like legends of some famous warrior, whenever I asked them directly, both Sensei would answer in the same standard way, "That was when I was young." Other than that, neither would say another word about it. This is the traditional way of the warrior class, essentially,

Jiko wo Kataranu
Never talk about yourself

I would like to take the time to introduce some of the teachers that I was recommended to train with. Kiyuna Sensei who's hands were so powerful he could, with just the palms of his hands, rip the bark off a tree. There was also Higaonna Sensei from Naha who looked like a Confucian scholar, along with Niigaki Sensei who I esteemed for his commonsense knowledge. Finally, I have to say the lessons taught to me by Matsumura Sensei, who many held in high regard, remain in my heart today.

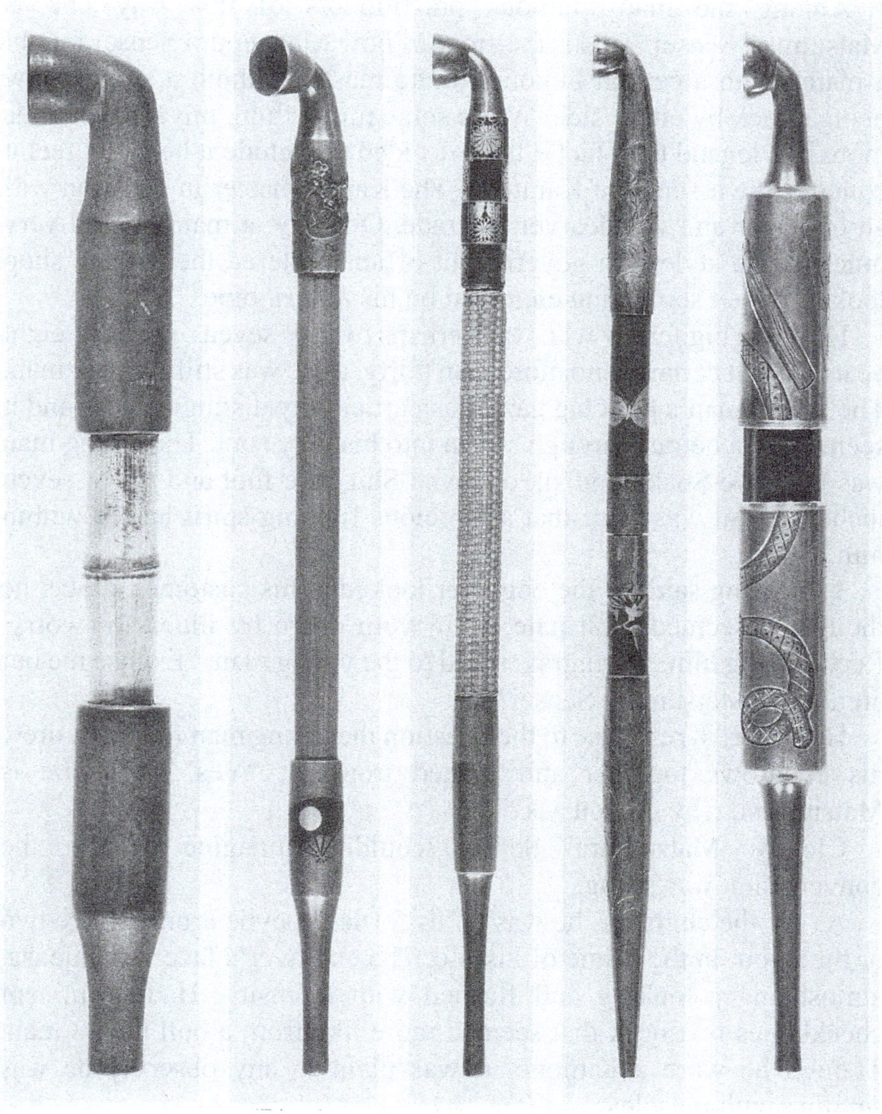

Examples of Edo Era Kiseru
From: *Kiseru Japanese Tobacco Pipes* 1988
By: The Tabacco and Salt Museum たばこと塩の博物館

A Tale of a Famous Duel

Among the many famous practitioners of this way one is Matsumura Sensei. This is the story of how Matsumura Sensei fought a match with a certain famous Karate master without a single blow being struck by either side. While some might think this story sounds more like legend than fact, I have decided to include it here as I feel it contains the essence of Karatedo. The Karate master in question was an engraver and woodcarver by trade. One day, a man dressed very much in the style of a government official entered the carvers shop looking to get something engraved on his *Kiseru* pipe.

He was a big fellow who was perhaps twenty-seven or twenty-eight years old but certainly no more than thirty, as he was still a young man. The young man's piercing gaze was particularly distinguishing, and it seemed to penetrate through a man into his very soul. The young man was tall, five Shaku and six or seven Sun, five foot and six or seven inches, and it was clear that a ferocious fighting spirit lurked within him.

That being said, as the engraver looked at his customer's face, he thought it seemed a bit pale, as if worn down by illness or worry. Recognizing him, the engraver said to the young man, "Excuse me but aren't you Matsumura Sensei?"

However, in response to the question the young man frowned, drew his eyebrows together and looked troubled. "Yes, my name is Matsumura, why do you ask?"

Clearly Matsumura Sensei couldn't imagine where the conversation was going.

As for the engraver, he was in his forties, maybe around forty-two or three, but in the prime of his life. The engraver's face was intense, almost angry looking and flushed with intensity. Had prominent cheekbones in a neck that seemed more like from a bull than a man. Though he wore a Kimono, it was plain to any observer he was rippling with muscles.

To distract the young Matsumura's puzzled gaze, the engraver reached out and took the offered pipe and, while fiddling with it in his hands continued, "I thought it was you. Truthfully, I have long hoped that if I ever met you, I could ask you to teach me."

Matsumura Sensei answered bluntly, "I regret to inform you that I do not instruct people in Karate."

Translator's Note: Lord of Ryukyu Domain

King Sho Tai 尚泰 (1843 ~ 1901)

Ryukyu existed as a Kingdom from 1469 until 1879 and was briefly a domain from 1872 to 1879 after which it became one of the forty-seven prefectures that make up Japan. The king in this era was Sho Tai, who transitioned from King of Ryukyu to "Lord (Governor) of Ryukyu Domain" until, in 1879, he was summoned to Tokyo and replaced with Nabeshima Naoyoshi, who became the first governor of Okinawa Prefecture.

However, the engraver did not relent, "Come now, if that is the case, who is the official Karate instructor to the head of Ryukyu Domain? There is hardly another Sensei as famous as you, or of your calibre!"

The engravers words clearly irritated Matsumura, who replied, "While it is true that I have offered my services to his lordship, I do not instruct others. In fact, I stopped teaching his lordship. There is one reason for this. I have come to despise Karate."

"That is a most astonishing thing to say. Matsumura Sensei, you are a close confidant of the lord of the domain himself! He has taken a liking to you, how is it you no longer enjoy Karate? Are you no longer serving as Karate instructor?"

Matsumura replied, "No, it is best that I no longer serve as official instructor because I was only able to teach an incomplete Karate. That is why I found myself unable to continue fulfilling the role of instructor."

"Now I'm even more confused!" the engraver declared, "If you, Matsumura Sensei, are not teaching, who is serving in the capacity of official instructor? This may seem a bit rude, but amongst the lord's retainers there is no one else I see being more capable than you."

"Indeed, it is for that reason I was sought out to take him on as a student. However, his lordship's technique is not very well developed. He needs to be spurred to further refine his skills. I could have allowed myself to be defeated, however that would hardly benefit his lordship. So, I challenged him directly asking if he thought he would be able to hit a living person with that level of technique. He became furious and launched *Nidan Keri*, double jumping kick. His kicks were very sharp. However, suddenly launching a double jumping kick at an opponent with superior skill showed his majesty's martial arts training was far from complete."

Matsumura continued, "In short, his attack was appalling and I thought I would teach him a lesson. After all a Karate duel is a duel to the death. If you don't understand that, in a duel, the first person to make a mistake will be the loser.... Then your training will go nowhere. You must enter a duel with the understanding that if you failed to stop your opponent's attacks, you will lose your life. Unfortunately, the Lord of the Ryukyu Domain did not approach training with that mindset. Thus, I wanted him to experience what would happen if you failed to attack or defend properly. So, when he launched a high kick flying towards my eye, I struck down hard with Shuto, knife hand, on my lord's foot. I then used my arm to block his second kick, hook his leg and sweep sending him flying. His majesty toppled down onto his side, and I slammed my body into him as hard as I could. That sent his body flying 2~3 Ken, 14~21 feet, across the floor." Matsumura concluded.

"Was his lordship injured?"

"He hit his shoulder hard and his hand was hurting. The spot on his foot where I struck him with a Shuto was swollen. It took him some time to get up as I recall."

"An astonishing story!... Did His Majesty scold you?"

"That goes without saying! I was ordered by His Majesty to leave and not return until summoned."

"I see, that is hardly unexpected. But at some point, he will no doubt forgive you for that incident?"

"I would say it seems unlikely. He was furious. It has been more than a hundred days since then and I haven't received the slightest communication. From what I've heard he's of the opinion that *Matsumura Sensei is extremely proud of what little technique he possesses*. It seems doubtful he will forgive me anytime soon. Thinking back, he would have been better off if I hadn't agreed to begin instructing him in Karate. No, even beyond that, I wish I had never learned Karate."

"That sounds very timid of you! Life is about being angry. Life is about laughing." The engraver then looked at Matsumura Sensei as if he were trying to console him and added, "If I might make a suggestion, teaching me may lighten your mood."

"I will absolutely not teach you! First, you are clearly the well-known Karate practitioner so-and-so, are you not? Why on earth would you seek out instruction from me at this point in your career?" Matsumura Sensei observed.[19]

"Truthfully, I do not require your instruction, I would simply like to see how a well-known Sensei teaches…"

While the engraver spoke to Matsumura Sensei politely in deference to the fact that Matsumura Sensei was born into the Samurai class and the engraver was from the Craftsman class, Matsumura Sensei nevertheless felt the older man's manner was somewhat condescending. This no doubt stemmed from the fact that the engraver

---

[19] Funakoshi Sensei does not list the engraver's name but uses the Kanji 某 that is used to refer to "a certain so and so" or "a certain person who shall remain nameless."

was a fairly adept Karate practitioner whose name was known in both Shuri and Naha. And no doubt he knew that the official instructor to the Lord of the domain could hardly start a fight, so he saw his chance to force Matsumura Sensei to teach him his methods.

At this point Matsumura Sensei was still a relatively young man and therefore engraver's condescending manner enraged him.

"Well, you are certainly persistent, but the answer is no, I will not teach you."

"I see, you will not honor me by teaching me?" The engraver next followed up with a disdainful glint in his eyes, "Then can I humbly ask you for a duel?"

"What do you mean a duel? Are you saying that you want a duel with me?"

The engraver now responded as if he was talking to a child. "That, good Sir, is exactly what I am asking for. In a duel no one is concerned with who is of high birth and who is of low birth.[20] Right now, if you will forgive me for saying this, but quite some time has passed since you were released from your official position as instructor to the Lord of the domain, and thus you do not need his majesty's permission. Trust me when I say I won't be kicking in a way that will mean my foot will swell up and I won't tumble in a way that hurts my hand. So, you don't have to worry about anything in that regard!"

"What's this?" Matsumura replied, "I have no idea how good you are, but I think your words are taking things too far. This is not a matter of getting hurt or not. In a duel you must be prepared to lose your life without complaint. Have you prepared yourself for that?"

"It is as you say. For my part, I am completely prepared." The engraver said mildly.

"Very well then, let's have a duel! When two tigers fight, one is injured and the other dies. It doesn't matter if you win or lose this duel you won't be returning home with your buddy unscathed. I will leave the decision regarding the time and place to you."

The pair agreed to meet at five in the morning on the following day at the graveyard of Kinbu Palace, which is behind Tama Palace.

20

Shiai ni Kisen no Betsu wa Arimasen 試合に貴賎の別はありません
In a duel no one is concerned with who is of high birth and who is of low birth.

So then on the day of the duel…Both men stood straight. They were about five or six Ken, 30~36 feet, apart. The engraver closed the distance with a series of short, quick steps that sounded like *ta,ta,ta,ta…*and stopped two or three Ken, 12~17 feet, from Matsumura Sensei, dropping his hips low. He held his left fist loosely in Gedan, Lower Stance, and kept his right fist on his hip, while staring intensely at his opponent, evaluating him.

Initially, Matsumura Sensei had been sitting on a boulder, but he now stood up. Though now standing, he hadn't altered his stance from that position. His left side was facing slightly towards the engraver, but the way his arms hung and his feet were positioned could only be described as a natural, everyday stance. Matsumura Sensei had, however, pulled his chin in tight against his left shoulder. His eyes had the same piercing stare described before, thought they were now opened even wider.

Seeing this, the engraver was slightly puzzled, and thought, "Is that really the stance he is going to fight in?"

The engraver couldn't understand why Matsumura, a well-known instructor, was standing that way. "Surely, he must be taking this duel seriously, whatever could he mean by taking such an awkward stance?"

The engraver continued to wonder, "Maybe he is simply overconfident? Or maybe he has simply gone soft in the head?" This last thought invigorated the engraver who decided that he would show his opponent something.

The engraver prepared to leap in with a ferocious assault, but the moment he was set to launch his attack, a bolt of lightning seemed to fly out of Matsumura Sensei's eyes, and, sensing this, the engraver immediately leapt back 2 Ken, 12 feet. Looking back at Matsumura Sensei, the engraver realized the older man hadn't moved a muscle.

Sweat was now pouring down the forehead and armpits of the engraver. His heart was pounding. The engraver suddenly sat down on the ground. For his part, Matsumoto Sensei simply sat down on the same boulder again as is nothing had happened.

The engraver thought, "I have no idea what just happened. Neither of us has struck a blow and yet I am drenched in sweat and I have no idea why my heart is beating out of my chest."

Just then Matsumura Sensei said, in a carefree voice, "The sun will soon rise, can we get on with it?"

Resolving himself, the engraver got to his feet. Matsumura rose up in a smooth motion as before, and simply looked at the engraver, without taking any stance whatsoever. This time the engraver positioned himself carefully before beginning to slowly inch forward. He closed the distance to first 5 Ken, 4 Ken, 3 Ken, 2 Ken. Thirty feet, 23 feet, 18 feet, 12 feet. The engraver was one breath away from the distance where he could strike, but found he couldn't move. It was as if his feet had been nailed to the ground.

Now, separated by the smallest of distances, both combatants had locked eyes and each man's fighting spirit roared across at the other.

Then, from the perspective of the engraver, Matsumura Sensei's fiery gaze seemed to take hold and bedevil him, and in that moment his fighting resolve evaporated. However, he could not tear his eyes away from Matsumura Sensei's gaze. He felt that if he diverted his eyes, even for a moment something terrible would happen. He had been cornered and had no chance to escape.[21]

Yaaa! The engraver shouted a ferocious Kiai.[22] Doing this should cause an opponent to respond in kind, and attack or perhaps respond

---

[21] *Zettai Zetsumei* 絶体絶命 desperate situation with no escape

[22] "*Kiai Jutsu* also known simply as Kiai, refers to using your spirit to pressure and control your opponent's spirit. A method for employing *Ishin Gekishiin*, to use your mind to attack your opponent. To put it simply, *Kiai Jutsu* means focusing your spirit and mental power on one point. This can be referred to as *Muichi Mukyu* Neither using anything or doing anything. It is also known as *Munen Muso*, Thinking of nothing, reacting to nothing, or being in a state of *Ishhin Furan*, Unified of mind without any distractions. You have determined your opponent's *Kyojitsu*, how he will feint and what his true attack will be. You are breathing in unison with your opponent using *A-Un* 阿吽 breathing which represent the first sound you make when you are born (A 阿)and the last sound you make when you die (Un 吽.) Kiai means using a shout embodying your total resolve that suppresses not only your opponent's fighting spirit, but disrupts his physical body, a power that can either kill or revive a person."

Konishi Yasuhiro (1893 ~1983)

*How to Become an Expert at Karat*e 空手上達法(1952)

According to the *Nihon Karate Do Shoto-kai* homepage, "Funakoshi Sensei used a long Kiai of *Hooii!* 「ほーいーっ」 Which was different from the typical shorter Kiai of Ei! 「えいっ」 or Yaa! 「やあっ」

with a Kiai of his own, and that would give the engraver a chance to pit a final technique against his opponent….however, Matsumura Sensei had not moved, despite the fact that the engraver's Kiai had thundered like an echo through a mountain valley.[23] It was as if it had not even registered in his opponent's ears. Matsumura Sensei simply stood there still and unmoving. The engraver leaped back.

"What is going on? I thought you were going to attack straight at me? Just shouting at me won't lead to victory!" Matsumura Sensei said as a smile played on his lips at the sight of his completely exhausted opponent.

"This is completely mysterious Sensei. I am not trying to brag, but up until now I have never had an opponent that truly challenged me."

"How do you think this bout will end?"

"…"

Matsumura's unnamed opponent did not respond but instead bowed his head and thought for a moment, before raising his head with renewed vigor and said,

"Let us continue, until one of us is the winner and one of us is the loser. No, that's not right, the bout has already been decided, no matter what I do from now on, I won't be able to recover my reputation. I will throw myself at you in a *Sutemi*, sacrifice attack, and see what happens!"

"Very well, launch your attack!"

Before the words "*Gomen*, forgive me!"[24] had even left the lips of the engraver, he had launched himself like a bullet.

In response Matsumura Sensei shouted *Gaa!* in a voice so loud, that to the engraver it was like a bolt of lightning crashing to the earth, hardly seeming a sound that could be produced by a human. The shout stopped him dead in his tracks. However, the engraver rallied his strength and tried one final time to charge, but the moment he met his opponent's eyes, he unconsciously gasped and fell on his face.

---

[23] *Kodama* 木霊 spirit dwelling in trees, but also refers to an echo. In the mountains, spirits responds when sound strikes them causing a sound to reverberate.

[24] This is the commonly known apology *Gomen* "I'm sorry" however in this case the meaning is *Burei wo Yuruse* 無礼を許せ forgive my lack of manners, which, in a combat situation, means "prepare yourself" similar to *Kakugo Shiro* 覚悟しろ "prepare yourself."

The wind blew Matsumura Sensei's hair wildly like it was the leaves of a Yomogi plant, just as the first rays of the morning sun began to rise in the east behind him. The golden light glittered in his hair, making seem like his head was wreathed in flame.

To the engraver, the fierce look in Matsumura Sensei's eyes, the earth-shattering shout erupting from his open mouth, made him seem like a powerful lord of light, scolding demons and evil spirits into submission through his great virtue. It was impossible for the engraver even to look at his foe. Struggling to get up onto his knees he said while placing the palms of both hands on the ground and bowing,

"I surrender, you have beaten me."

"You are surrendering? Mr. So-and-so, famed for his Karate skill is saying this!?"

"I beg your pardon, the duel was a total disaster. I am now keenly aware of how amateurish my technique is. I am deeply ashamed that I allowed myself to be called Mr. Karate So-and-so."

Matsumura Sensei held his hand up to stop the man from further reproaching himself,

"No, that is not true at all. You not only possess a powerful fighting spirit and also have admirable technical ability. If your technique was pitted against mine, I doubt I would emerge the winner."

"How could that be? I was unable to use my arms or my feet? Sensei, your gaze was so penetrating and your face so frightening, it was as if all my will to confront you as an enemy flew away!"

Faced with a bewildered opponent, Matsumura Sensei spoke quietly saying,

"I have no doubt that is how you felt. You approached this duel thinking that you would win. I entered this duel seeing it as an enjoyable way to travel down a path that lead to my death. That is the difference between us. Yesterday when you invited me to duel with you, I was twisted up with worry, wondering how angry the lord of the domain get at me. In short, I was pondering this and that. However, as it turned out, the moment I decided to engage you in a duel, all that doubt and worry quite mysteriously evaporated."

Matsumura Sensei continued, "Thinking back, I realized that I had become too attached to things. I was attached to Karate techniques, attached to teaching Karate, attached to concerning myself with the lord of the domain's moods. To put it simply I was obsessed with my personal circumstances. Humans only exist temporarily as a manifestation of the five aggregates, matter, sensation, perception, mental formations and consciousness[25] and the five elements, earth, water, fire, wind and void. The moment our lives are extinguished we return to the five elements: earth, water, fire, wind and void. Recalling that the Buddha teaches *Form is emptiness and all is vanity*[26] I realized that if I do not exist, then neither does my opponent. There is no death. If you have no attachment to anything, then all obstacles before you are swept away…having realized that, everything became clear."

Later when the engraver told other people this story, he always praised his opponent by saying, "Matsumurua Sensei is the very definition of a great person."

It goes without saying that Matsumura Sensei once again became the official Karate instructor to the lord of the domain.

The Secret Martial Art

Where did Karate originate? Who invented it? How did it develop? Unfortunately, as there are no historical documents related to this the answers remain unknown. Our only source of information are the few oral traditions and legends that have been handed down. Further, those oral traditions are extremely fragmented and vague.

The fact is, even in the years following the grand Meiji Restoration in 1868, I and other youths were unable to practice this art publicly. In other words, even in the late 19th century, the art of Karate was clouded in secrecy.

In that era there were no Dojo. Thus, training was simply an instructor and his students. Of course, such Sensei did not teach Karate as a living, but had another profession. Those that taught Karate only did so as a hobby and those that studied it only did so because they enjoyed it. For example, I was Asato Sensei's only student while Itosu Sensei only had a handful of students. It seems likely the reason not a lot of emphasis was placed on written records, and other such resources was because there were no teachers that taught Karate as a profession.

As a person seeking to pass on Karate to future generations, I feel this lack of written information is unfortunate, however it can't be

---

[25] *Goun* 五蘊 a Buddhist term referring to the five Skandhas (the components that make up our individual self or being.)

[26] *Shikisoku Zeku* 色即是空 form is emptiness, matter is void, all is vanity.

helped. However, I have committed myself to leaving a record of what my teachers taught me or the answers they gave to my questions, though I must admit to being concerned about whether I may have misheard something or if my memory is accurate. Be that as it may, I plan to simply introduce the teachings handed down to me as well as what I remember.

There is a legendary story from long ago regarding Napoleon. He heard there was a small independent nation in the East where nobody carried any weapons. This clearly referring to Okinawa Prefecture, which, at that time, was known as Ryukyu. [27]

Generally speaking, Karate is thought to have flourished after the government passed decrees banning weapons as part of their government policy to demilitarize the nation.[28] There were two of

---

27 "Interestingly, Hall discussed Okinawa with Napoleon himself when the ship Lyra put in at St. Helena, (where the now former Emperor now lived in exile) and reported in his account that:

> Several circumstances…respecting the Loo-Choo (Ryukyu) people surprised even him a good deal; and I had the satisfaction of seeing him more than once completely perplexed, and unable to account for the phenomena I related. Nothing struck him so much as their having no arms. "Point d'armes!" he exclaimed; . . . "Mais, sans armes, comment se bat-on?"
>
> I could only reply, that as far as we had been able to discover, they had never had any war, but remained in a state of internal and external peace. "No wars!" cried he, with a scornful and incredulous expression, as if the existence of any people under the sun without wars was a monstrous anomaly."

Okinawa: The History of an Island People
George H. Kerr
1958

With the exception of a momentary fit of scorn and incredulity when told that the Loo-Chooans had no wars or weapons of destruction, he (Napolean) *was in high good humour while examining me on these topics.*

-*Voyage to Loo-choo* (Ryukyu)
By Basil Hall (1788~1844)

28 *Kinbu Seiji* 禁武政治 Official government policy of banning weapons.

these decrees, one about five hundred years ago and another two hundred years ago.

The first was just after the main island of Okinawa was unified. Just before this, the main island of Okinawa was divided into three kingdoms that were constantly at war with one another: The Central Mountain Kingdom, the Southern Mountain Kingdom and the Northern Mountain Kingdom. Into this conflict an extraordinary man of great bravery emerged. He was Sho Hashi, the King of the Middle Mountain Kingdom and he unified all the Three Mountain Kingdoms under his rule. Having gained control of the entire nation he instituted a strict demilitarization policy.

The ban on weapons meant you couldn't even walk around with a rusty old sword. Further, government officials and academics from all the Three Mountain kingdoms were ordered to gather in the capital, meaning King Sho Hashi centralized power in Shuri. Following a shorter rebellion and restoration, the later Kings of Ryukyu, the Sho Clan, and their descendants ended the conflict between the Three Mountain Kingdoms, that had led to century-long warring states period in Ryukyu and established peace over a unified island.

That peace lasted for over two hundred years. However, in the 14th year of Keicho (1609) the Ryukyu Kingdom was forced to battle Satsuma domain, the fierce southern-most domain in Japan, led by the Shimazu clan.

At that time, the Shimazu Clan was led by Satsuma Hayato, a Samurai of dauntless courage, whose prowess as a taction was feared by all the neighboring domains.

This invasion awoke the fighting spirit of the Ryukyu warriors and, initially, their ferocious attacks frustrated the Satsuma invaders, however a surprise attack by a special Satsuma attack squad landed ashore unexpectedly and the tide of battle reversed. In the end Ryukyu surrendered to Shimazu forces.

Having lost to the invaders, the Kingdom of Ryukyu fell under control of the Shimazu Clan. From that point on, possession of weapons was banned. It seems likely that in response to this ban, members of the warrior caste, began to conduct training in secret, which lead to the rapid development of an unarmed method of combat.

Even before these events, as there was a great deal of trade between Fuzhou Provence and Ryukyu, Chinese-style Kenpo was transmitted to Ryukyu. This Kenpo either merged with the unarmed fighting art native to Ryukyu or was practiced independently. It is thought this is the origin of *Okinawa-te*, Okinawan Martial Arts, and *Tode*, Chinese Martial Arts, the precursors of Karate.

When I was young, I would often hear old men talking about *Okinawa-te* or *Kara-te*, however it seemed to me that when people spoke of *Okinawa-te*, they were referring to the martial art unique to Okinawa while *Kara-te* referred to Chinese Kenpo.

Thus, because there were two governments that passed military disarmament decrees, it is quite natural that the people developed a method of self-defense that was *Toshu Kuken*, Barehanded and Unarmed, to topple opponents. As the Shimazu Clan sent domain policemen to ensure their rules were being followed, it is hardly mysterious that training such an art and transmitting its teachings was done in secret.[29]

Even when I began training, lessons had to be taken secretly at night. Clearly this is a testament to the deeply ingrained restraint that was a holdover from the days when Ryukyu was under Satsuma rule.

This is just my speculation, but I believe training was concealed as dancing in order to deceive the eyes of the Daimyo in Satsuma. If you look at the Okinawan dances practiced today in places like Tokyo or Osaka, and compare them to traditional mainland dancing, you will notice that while the latter is very fluid and graceful arm movements, the former very clearly focuses on moving from the hips.

Taking a closer look, you will note that whether it is men dancing or women, the feet movements, hand movements along with powerful hip thrusts, each movement is clearly heavily influenced by Karate. Sometimes there are even parts that seem to be exact copies of Karate techniques.

Indeed, the essence of Okinawan dancing is the graceful way dancers begin and end performances. Looking at this aspect, it is impossible to say there is no connection to Karate. This is perfectly in line with the philosophy of, "Karate begins and ends with a bow of respect."

This type of overlap is something that happens in Okinawa. Further, the people of Okinawa consider their country *Shurei no Kuni*, a Land

---

[29] In 1628 Satsuma Domain established a *Bugyo*, police station, in Naha. It was staffed by about twenty Samurai from Satsuma Domain for the next two-hundred fifty years, until 1872 when Ryukyu became a domain under the Meiji government.

of People Who Strictly Follow Proper Etiquette. A country that is governed on the fundamental principle of courtesy to others.

Thus, up until the outbreak of the war, the main gate to Shuri Castle had gate known as the *Shurei Mon*, Gate of Revering Etiquette. In fact, there was a large, engraved sign that said, *Shurei no Kuni*, Land of People Who Strictly Follow Proper Etiquette.

It goes without saying that there are no documents affirming the connection between *Land of People Who Strictly Follow Proper Etiquette* and the martial art Karatedo that *Begins and Ends with a Bow of Respect.*

Following the first abolishment of domains and replacement with prefectures by the Meiji government, Shuri castle was designated a National Treasure along with the *Shurei Mon* gate. However during the war, it was blown to smithereens.[30] Nowadays the space is occupied by an American military base, that shows no sign of removing itself. It is ironic that *Shurei*, which even now continues to be a symbol of the people's desire for democracy and peace, is now being piteously trampled under an ironclad rule

[30] *Unsan Musho* 雲散霧消 leaving nothing behind like clouds or mist after they have departed.

How Karate, Chinese Hand, became Karate, Empty Hand

Of late most people use the Kanji "Empty Hand" when writing about Karate however, some people still ask me, "The word Karate, Empty Hand, refers to Karate, Chinese Hand right?"

To be fair, using the Kanji "Empty Hand" to write Karate is a recent affair that only began after I began teaching in the capital. So, while it is undoubtably a new way to write the word, this does not mean that writing Karate with the Kanji "Chinese Hand" is correct.

It goes without saying that the word Karate is used in Okinawa to refer to this art. In this case *Kara*, with no defined meaning, is written in Kana and *Te*, hand, is written with one Kanji. However, typically Karate is simply referred to as *Te*, meaning "Martial Arts." It was also known as *Bushi no Te*, the Way of the Warrior. Thus, people would say things like,

Ano Hito wa Te wo Shitteiru soda!
That person is knowledgeable about Karate!

Te wo Shugyo Shita soda!
That person has trained in Karate!

Bushi no Te wo Shitteiru!
That person knows about the Way of the Warrior!

Gradually the use of the word "Karate" increased, though it is unclear whether the first Kanji was "Chinese" or "Empty." As there are no historical documents related to this, it is impossible to give a definitive answer.[31]

As I previously mentioned, writing Karate as Chinese Hand, implies that the art was the *Tode* style of fighting transmitted from China. However, this probably had more to do with the fact that at the time in Ryukyu, there was a great deal of excitement, bordering on reverence, over anything Chinese.

---

[31] Funakoshi Sensei is indicating that "Kara" is a native Ryukyu word that does not refer to either China or the word Empty.

Members of the Keio University Karate "Chinese Hand" Club
1920s

Thus, the style at the time was to attach the Kanji *Kara* meaning "Chinese" to any and everything. In short, it is not clear how the Kanji *Kara*, the first Kanji in *Tode*, became associated with the word Karate, which up until now had been written with Kara in Hiragana and *Te* in Kanji, but it seems to me that since the first part of *Kara-t*e happened to coincide with Kara "Chinese" so at some point that Kanji began to be used as it fit so perfectly.

Kara	唐	China/ Chinese
Tode	唐手	Chinese Hand
Karate	から手	Kara Hand
Karate	唐手	Chinese Hand

To summarize, it would not be an exaggeration to say that historically, in Okinawa, martial arts were simply called *Te*, Hand. Further, *Te* was primarily used with terms such as *Shuri Te* or *Naha Te*.

In all likelihood, these various names are probably associated with certain schools of martial arts and the term *Te* probably more accurately describes the martial arts practiced at the time than *Kara-Te*, or Karate written with the first part in Kana alphabet and *Te* in Kanji, however as there are no written records, it is impossible to answer definitively.

Therefore, as writing the word Karate with the Kanji "Chinese Hand" became common, a person unfamiliar with this art could easily mistakenly assume this was a Chinese style of Kenpo based on the way the word is written. In fact, even today there are more than a few people that believe this to be so. To be clear, the Kata and Kumite that so many students studying today, along with the training methodologies, are unique to Okinawa and differ markedly from those in Chinese Kenpo.

As you can see from the above passage, I too have some concerns about the word Karate when written with the Kanji "Chinese Hand." This came to a head a few years after I traveled to the capital in the early Showa Era. [32] The Karate "Chinese Hand" club had already been established at Keio university and one time I was talking with the

[32] The Showa Era began in 1926 and the "early Showa Era" is generally from 1926 until 1935.

students and we decided to change the name to the *All Japan Kenpo Karatedo*, using the Kanji "Empty Hand" for Karate.[33]

At the time there were some people in both Okinawa and in Tokyo who were heavily critical of this change, however I was confident in my decision. So, in the end, I was able to push the change through, and it succeeded to the degree that if today someone were to write the word Karate using the Kanji "Chinese Hand" then most people would see it as strange.

---

[33] Fujita Seiko (1898 ~1966) was a Japanese martial artist and martial arts researcher who was friends with many martial artists including Mabuni Kenwa. He wrote numerous books about traditional Japanese martial arts and Ninjutsu. Fujita wrote the following regarding using the Kanji "Empty Hand" for Karate:

During the following Showa Era (1926~1989) I attended a conference related to martial arts. While chatting with some people I said the following:

Kenpo is Toshu-Kuken, Fighting Barehanded Without Weapons. Both Karate (Chinese Hand) and Kenpo (Fisticuffs) use these unarmed techniques to topple the enemy. So rather than using the Kanji 唐 *"Kara" representing China, we should be using the Kanji "Kara"* 空 *representing Empty.*

A good friend mine who was sitting at the same table, was a Karate instructor, nodded in agreement and said,

That sounds about right to me!

After that Karate 唐手 (Chinese Hand) was changed to Karate 空手 (Empty Hand.) That is how we ended up with the techniques of Karate being known as the Empty Hand style of fighting.

-The Eighteen Weapons of War
Fujita Seiko
1958

As for the Kanji *Ku*, it represents the state of mind in martial arts characterized by the phrase,

Tote Kuken nishite Mi wo Mamori Teki wo Fusegu
徒手空拳にして身を守り敵を防ぐ
Protecting your body and defending against an enemy while weaponless and empty handed

Further, any person following the path of martial arts will,

Kukoku no Koe wo Tsutaeru ga Gotoku
空谷の声を伝えるが如く
Echo like a shout into an empty valley

By doing this you rid yourself of selfish intent and polluting thoughts thereby achieving a state of,

Shinchu Kukyo
心中空虚
Completely empty in mind and completely calm in body

Which is essential.[34]

---

[34] *Anyone studying Karate "Empty Hand" should reflect like a polished mirror or echo like a shout into an empty valley they have been stripped of intent and polluting thoughts, completely empty in mind and completely calm in body, waiting solely to acquire the skills to achieve the ultimate level.*

-Funakoshi Gichin as quoted by Nakayama Masatoshi 中山正敏 (1913~1987)
Karatedo Instructors Course 空手道新教程
1965

You must be,

Chuku nishite Gaichoku
中空にして外直

"Hollow on the inside, straight on the outside" A mind free of polluting thoughts, who approaches life with solid principles. [35]

Also, Buddhism teaches,

空即是色, 色即是空
Kusoku-zeshiki, Shikisoku-zeku
Form is emptiness, emptiness is form. [36]

All things and phenomenon in our realm that you observe are voids, but if you come to realize and understand things are formless, those very voids become real. There are a great variety of martial arts such as Jujutsu, sword fighting, spear fighting, archery and staff fighting, however if you examine them closely you will find their underlying principles are the same as those of Karate. The truth is, Karate is absolutely a fundamental art, which is why I insist upon using the Kanji *Kara*, meaning void.

While I would like to expand about my philosophy regarding using the Kanji *Kara*, I have limited space in this book, so the above simple explanation will have to suffice. I plan to expand on this topic in my forthcoming book *Karate Kyohan, An Instructor's Guide to Karate*.[37]

So then, the way Karate was written changed from "Chinese Hand" to "Empty Hand" and our organization was now called the *All Japan Kenpo Karatedo Research Society Headquarters*. I could not ignore the fact that each of the techniques needed to be improved. First, many of the names of the techniques were either in Okinawan language or were heavily influenced by Chinese and were therefore difficult to understand.

---

[35] The phrase equates the stem of a lotus, which is hollow and grows straight to the ideal man of virtue. Attributed to the Chinese philosopher Zhou Dunyi 周敦頤 (1017~1073.)

[36] This is a line from the *Heart Sutra,* which is a popular sutra, collection of aphorisms, in Mahayana Buddhism.

[37] Referring to a planned expanded edition.

Translator's Note: Evolution of the word Karate to "Empty Hand"

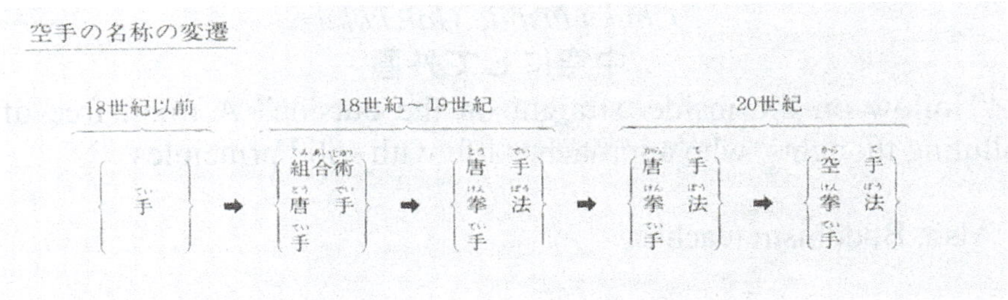

Pre-18th century
Tei "Hand"

↓

18th ~19th Century	
Kumiai Jutsu "Paired Training" *Totei* "Chinese Hand" *Tei* "Hand"	→ *Totei* "Chinese Hand" Kenpo "Way of the Fist" *Tei* "Hand"

↓

20th century	
Karate "Chinese Hand" Kenpo "Way of the Fist" *Tei* "Hand"	→ Karate "Empty Hand" Kenpo "Way of the Fist" *Tei* "Hand"

-Explanation of Okinawan Karatedo: History and Techniques
精説沖縄空手道：その歴史と技法
Uechiryu Karatedo Organization 1977

I felt that leaving the Kata unchanged would make them difficult to learn for people who lived on the mainland, where people live a different lifestyle than in Okinawa. In addition, I also wanted Karate to spread as a form of physical fitness, so I simplified all the techniques.

Things change and this doesn't apply just to martial arts. The Karate that the young people practice in school clubs, doesn't look anything like the Karate I did a decade ago. And that Karate doesn't resemble the Karate we were taught when people of my generation started. There were never any troublesome and hard fast rules regarding Karate Kata when I first started training, no one ever said,

Kaku Araneba Naranu
It must be done this way[38]

Further, since no such admonitions exist today, as long as you are maintaining the traditional *Seishin*, mindset, and using *Seiken*, a proper punch, it is quite natural that your method will differ to a degree from another person's and your approach to a given situation will reflect that.[39] In the end, my goal is to make Karate a method of physical education that can be easily practice by anyone, young or old, man or woman.

In the 11th year of Taisho (1922,) just after I moved to Tokyo the publisher Bukyo released my book *Karate: Ryukyu Kenpo*. At the time, Karate, Empty Hand, was not something the general public was familiar with, so to ensure that readers understood what the book was about, the descriptive title of *Chinese Hand Boxing of the Ryukyu Kingdom* was used. However, in the book I listed the techniques using the Okinawan language names. For example,

Pinan, Naihanchi, Chinto, Bassai, Se-shan, Jitte, Jion, Sanchin.

These are the names I learned when I was first training under my teachers but there was no information on who invented these techniques or any other historical information. However, these names are not suitable for a general audience, so around the same time I

---

[38] This is an old-fashioned way of speaking Japanese, closer to the Edo era than modern times.

[39] Funakoshi Sensei seems to be referring to the strike Seiken as well as the meaning of the Kanji 正拳 True + Fist.

switched the way the word “Karate” was written from “Chinese Hand” to “Empty Hand” I renamed the techniques as follows:
Ten no Kata, Chi no Kata, Jin no Kata, Enpi, Gankasku, Hangetsu, Meikyo, Hakko, Kiun, Shoto, Shoin, Hotaku, Shokyo and so on.

However, I do not claim that these names are definitive and must remain so. If, a few years from now, it becomes necessary to change the names, I have no issue with that. I will entrust those training at the time to make that decision.

Translator's Note: Pronunciation and Kanji

Okinawan Pronunciation	Kanji used in the Shotokan	Other Kanji combinations
ワンシュー Wanshu	燕飛 Flying Swallow	腕秀、王輯、王集
チントー Chintoo	岩鶴 Stone Crane	陣闘、陣東、鎮党
ジッテ Jitte	十手 Police Truncheon	十手、実手、術手
セーシャン Seeshan	半月 Half Moon	十相
ジオン Jion	慈恩 compassion	慈恩、慈音
ミョウキョウ Myokyo	明鏡 Bright Mirror	n/a
ソーチン Soochin	八荒 Everything	壮鎮、壮陣
ニーセーシー Nii See Shii	撥雲 Striking Cloud	二十四
ショウトウ Shoto	松濤 Shoto	n/a
チンテー Chintee	松陰 Shadow of the Pine	陣手、珍手、鎮定、鎮手
ゴジューシホウ Gojuu Shiho	鳳啄 Woodpecker	五十四歩
ジイン Jiin	松喬 Pine Bridge	慈蔭、慈允

Karate Schools

There has been a lot of discussion of late regarding the different schools teaching this art. Generally speaking, having schools branch off in many different directions will not benefit Karatedo as a whole.

It is said that the oldest schools in Okinawa are *Naha-de* and *Shuri-de*.[40] Further, it is said that these two styles are related to Wudang Mountain School and the Shorin School from China. It is thought that the Kata found in Okinawa were created through a complicated interaction between various schools and sub-schools of martial arts.

In China, during the time spanning the Yuan, Ming and Qing dynasties (1271~1912) the Wudang Mountain School separated from the Shorin School and the two engaged in *Sessa-Takuma*,[41] an intense, but friendly rivalry that leads to both schools excelling. In the end the two schools can be described as *Iccho Ittan*,[42] each having its strong points and weaknesses.

The founder of the Wudang Mountain School was a man named Chang San-Feng. At first glance the strikes in this school seem soft and completely lacking in power, however when seen it in action, you realize that one blow can easily topple a person.

There is one theory that states the Shorinji School was founded by the Great Teacher Daruma. Shorinji teaches various methods to strike with the hands and feet. This school has many approaches to teaching techniques, including both a hard and a soft method.

Further, both the above mentioned schools were practiced by all strata of society in China up until the end of the Qing Dynasty. Tehre were frequent public demonstrations of this traditional art and it seems many Chinese proudly proclaim it to be the national martial art of their country.

The Wudang Mountain School takes its name from the nearby Wudang Mountain where it was developed. The Shorinji School takes its name from the Shorinji Temple in Henan Domain. The Great Teacher Daruma travelled from "a great distance", in other words from India, across mountains and grasslands until he reached China. When he began teaching Buddhism at the Shorinji Temple in Henan, he was appalled to find that students attending his lectures were

---

[40] The reading given is "De" instead of "Te."

[41] *Sessa-Takuma* 切磋琢磨

[42] *Iccho Ittan* 一長一短

collapsing and unable to continue. Daruma introduced his new method which not only trained the mind but the physical body as well, saying,

> *Generally speaking, the teachings of Buddha are relayed to instruct the spirit, however your spirit and body are fundamentally unified, and cannot be separated from each other. From what I can see, you all are exhausted not only in mind but also in body, meaning that it is highly unlikely you will be able fully complete your training. Starting tomorrow morning, we will rise early to study the complete lessons imparted by the Buddha.*

At some point this method evolved into a *Taijutsu,* method of moving your body and manipulating the body of your opponent, and the name Shorinji School became associated with it.

This ends this topic however, I cannot ascertain how much of these origin stories are true and how much are simply legend. Nevertheless, a *Kenpo no Yona mono,* Boxing-like Art, was developed in China and eventually transmitted to Ryukyu. It is thought that it merged with the native martial art of Ryukyu and formed the base of what we know today as Karatedo.

Initially, Karate was taught in two different styles, the Shorin School and the Shorei School. However, it is not clear if these styles are related to the Shorinji School that developed in China and how they are related to Shuri-de and Naha-de remains unclear.

Generally speaking, Shorei style martial arts are best suited to larger, stronger men with thicker bodies while Shorin style is the opposite, suited towards men who are thinner and lack power. That being said, each style has its advantages and disadvantages.[43] While the fundamental stance used in Shorei School is solid, in practice such Shorei practitioners lack the ability to respond nimbly. While Shorin practitioners can move nimbly, once they have moved in close to finish an opponent, they are in danger as their technique may become ineffective at that range. Thus, since both schools have advantages and disadvantages, it is necessary to merge the two styles. This is the spirit of what Karatedo does today, incorporating techniques from each school.

---

[43] Funakoshi Sensei again uses *Iccho Ittan* 一長一短 Some strengths, some weaknesses.

In short, I believe it is important in Karate to get away from the concept of having different *Ryugi,* or schools. As an example, there are certain Shihan, head instructors, that have a particular affinity for one Kata or another, and focus on it exclusively, or there are those that publicly declare, in detail, the fact that they are such-and-such branch of this-or-that school. Ideally, to enhance the promulgation of this art, I feel they should all use the term Karatedo.

There are people who describe students in my Dojo as practitioners of "the Shotokan School," however, as was previously mentioned, I cannot agree with that description.

My Wife's Karate

I previously described how my ancestors were hereditary lower-class Samurai. My grandfather, Gifuku, was a well-respected Confucian scholar and esteemed by the family that controlled Ryukyu domain.[44]

He worked at the Kikoe Ogimi Udon, which was the temple and residence of the Kikoe Ogimi, the highest-ranking priestesses of the Ryukyu Kingdom, who venerated the ancestors of the royal family. The priestesses were the widows of the lord of the domain or widowed daughters of members of the imperial family. Men were forbidden to enter. Be that as it may, my grandfather was the type of person who members of the Samurai class would take three paces forward to meet.[45] He was fawned over due to his "poor but honest" style.

However, when grandfather Gifuku began to get on in years, he requested that he be allowed to retire from service. The former king, now lord of Ryukyu domain, granted him a residence in Teracho Town near the temple residence. Just as the great Meiji restoration was occurring my grandfather received a generous pension, however my father's generation lost it all.

---

[44] Ryukyu was an independent kingdom up until 1872, whereupon it became a domain under Japan, with the Ryukyu King as its head. Ryukyu Domain was then changed to Okinawa Prefecture in 1879 and a governor replaced the King, who lived in Tokyo until his death.

[45] *Demukae Sanpo, Miokuri Nanaho* 出迎え三歩、見送り七歩 This refers to the polite way to greet guests. When a guest arrives, you take three steps out to greet them. When it is time for them to depart, walk with them seven steps.

As for my father, unlike me, he was tall and handsome and quite skilled at Bojutsu. Unfortunately, he was addicted to singing and dancing and was a well-known drinker. Thus, those habits were probably the reason my father had to give up the house my grandfather received. In the end, we grew up in a rented house.

As I was from a poor household, I did not get married until what was considered at the time quite late. At any rate, from the day I brought my bride into our household I alone was responsible for supporting not only my wife but both my parents and grandparents. A large family indeed.

At the time my salary was ¥3 a month.[46] Nowadays, some people work part-time jobs to supplement their income, but there was nothing like that at the time. Further, a person working as a teacher would never be allowed to do work on the side. At the same time, I did have another important job, one that did not even earn me one Sen, the work of learning everything about Karate.[47] Thinking back on it, the fact that I was able to support a family of almost ten people while developing my skill in Karatedo is in large part due to the diligence of my wife.

Despite all this there were times when we didn't have enough money to pay the 25 Sen rent on our house. Further, when festival days came around, we were always hard pressed to come up with funds to buy offerings.

Therefore, to supplement our income, my wife would rise before dawn and walk half a mile to a small plot of land we farmed. She would also work late into the night weaving the famous Ryukyu Kasuri. At the time one bolt of this fabric would bring in 6 Sen, which was a pittance comparted to the work involved. However, my wife kept at it as the funds helped our financial situation. Her fortitude was a wonder to behold but I often wondered when she actually slept.

---

[46] At that time on mainland Japan, a ten-kilogram bag of rice was about ¥1, a plate of curry and rice was ¥0.07 and a bottle of beer was ¥0.17. However, prices changed dramatically in the Meiji era. For example, in 1868 the same ten-kilogram bag of rice was ¥0.55.

[47] One Sen was 1/100th of a yen. At the time you could buy a sweet-bean doughnut for 1 Sen.

ミンタリー笠

Mintari Gasa
Mintari Hat

I would also sometimes work on our plot of land, however as I was a schoolteacher, being seen working in the fields with my wife would have caused a stir, so I was sure to always wear a *Mentari*, woven straw hat, when I set out to do farming.

Not only was my wife was always encouraging with regards to my Karate training, she also never complained about the time I spend in that pursuit. Interestingly, my wife took to imitating my Karate movements and, at some point, seemed to have developed skill in this art. She was also a very insightful regarding Karate.

Thus, while for all appearances she was a typical housewife, when faced with a mountain of chores to do, instead of asking her children to drum their fists on her neck and shoulders or have them rub her lower back to relive stress, my wife would drill several Karate techniques to refresh herself.

After I noticed she was doing this, I made a point of observing her, but once she had begun, she continued training Karate as another part of her daily life.

Around that same time, when I couldn't train with Asato Sensei or Itosu Sensei, I would train by myself in the backyard of my house. At some point, first one, then two people began to show up as they were interested in Karate. Before long a great many young Karate practitioners were showing up at my house. If I happened to be late returning home, my wife would address any students that had already assembled,

"He's going to be a little late today, but that doesn't mean you should just stand around doing nothing until he gets home, go ahead and get started!"

Then she would stand and watch the practice and even occasionally correct mistakes,

"That isn't correct, use your left…Left!"

Sometimes she would go so far as to actually demonstrate the correct way to do a Kata.

"That is not how it is supposed to go! Your right fist should be doing this, alright?

Translator's Note: Various Tales From the Southern Isles

In the 1850s a Samurai named Nagoya Sagenta was banished to Amami Oshima for participating in and family succession revolt. (The other participants were ordered to commit Seppuku.) While there he recorded his observations about Ryukyu life in a book titled **南島雑話** *Various Tales from the Southern Isles.* The term "Southern Isles" was a common way to refer to Okinawa well into the early 20th century. There are many illustrations of Ryukyuans drinking and fighting.

Furious woman with an axe

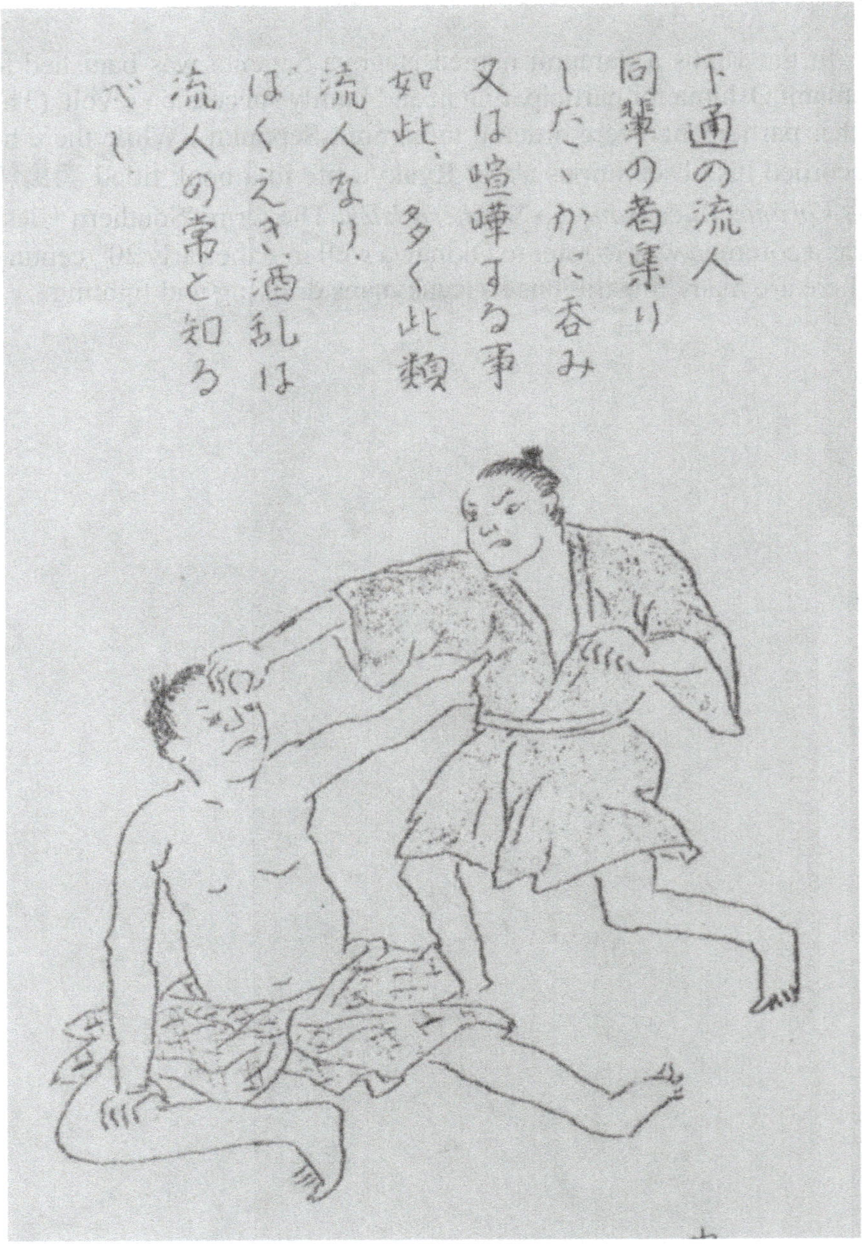

Whenever lower class Ryukyu people gather to drink, fights invariably break out as shown in this illustration. Many Ryukyu people are like this. Know that many Ryukyu people are heavy drinkers.

Thus, our house was frequently packed with people and rather noisy, however my neighbors were Rickshaw drivers or Tofu makers (Tofu makers in Okinawa are not like the ones in Tokyo that you find along the main street, in Okinawa Tofu shops can only be found on back streets, where they conduct business from small shops)[48] along with makers of Sukigushi combs or Waraji straw sandal shops. These people often got into arguments or fights. If yesterday it was kids brawling with each other, today it would be a bunch of drunks falling all over each other.

I was generally unaffected by this as during the day I was at school and at night I would be training with Asato Sensei or Itosu Sensei, but I heard there were some spectacular fights. Invariably it would be my wife who would mediate the situation.

It goes without saying that your average woman would not try to intercede between two men who have gone from arguing to grabbing each other's collar to a full-blown fist fight. Even a man would not want to get in between two men engaged in some tedious dispute, however my wife was able to slip in and ameliorate the situation.

It goes without saying that she did not use violence. Amusingly, with time, even the local youths admitted my wife had a way with words and could be relied on to defuse situations.

Something Long Hidden, Revealed For the First Time

As far as I know it never happened before around the 34th or 35th year of the Meiji Emperor (1901~02.)

It happened when Ogawa Ginnosuke, the prefectural school inspector, made the trip from Kagoshima to Okinawa. During the meeting my revered teacher Itosu Sensei was in attendance and he demonstrated Karate Kata in front of the inspector. After the demonstration, Mr. Ogawa, who was obviously impressed, asked Itosu Sensei many questions about Karate before he departed. This was a turning point as shortly thereafter the Number One Prefectural Boys Middle School and Boy's Teacher Training Schools opened with Karate being taught as part of the physical education curriculum. I later heard that Mr. Ogawa raved about the refined excellence he could see in each part of the Karate demonstration to the Ministry of Education.

---

[48] Brackets are by Funakoshi Sensei.

It is difficult to describe the intense emotions I felt at that time.

When I began Karate, training was basically done in secret. Now, I was in the midst of seeking the inner mysteries of the art while living a life of toil and poverty, when suddenly, on the day the Ministry of Education recognized this art, my outlook cleared and I became able to both teach and train Karate at school. It goes without saying that this reinvigorated my commitment to Karate.

Due to this announcement, Karate began to flower and bloom all over Okinawa and began to flourish. Karate instruction began at all Junior High Schools as well as youth clubs and even elementary schools.

I had, at long last, become a consultant regarding Karatedo training and instruction. I remember the first time I stood in front of a class of children, having received the approval of both Asato Sensei and Itosu Sensei, I was so overjoyed that I felt like this is what I had been born to do.

Several years later, Captain Yashiro Rokuro, who would later become admiral, entered port with his training fleet.

He attended a demonstration of Karate conducted by the children and was extremely impressed. So much so that he immediately ordered all his young officers and petty officers to line up and undergo Karate instruction. It seems likely that this is the first time the military encountered Karate.

Later, in the first year of Taisho (1912) the Imperial First Fleet arrived in Naka Gusuku Bay under the command Admiral Dewa. The admiral selected a dozen or so non-commissioned officers and sailors and sent them to the Number One Prefectural Junior High School. They stayed for a week, training Karate every day.

Thus, little by little, mainland Japan became aware of the existence of this art, some of which was due to simple word of mouth. However, it took ten years from that point until a Karate practitioner introduced Karate on the mainland.

When I say "ten years" I am referring to the tenth year of Taisho (1921) when his majesty the crown prince (now the Showa Emperor) paid a visit to the southernmost part of Japan, Okinawa, before embarking on a tour of Europe. We were honored to have him attend a demonstration of Karate.

首里城　龍樋

Dragon spout at Shuri Castle
Showa Era Picture

The man in charge of the destroyer the prince was travelling on was Captain Kanna Kenwa.[49] I believe it was Captain Kanna, who later became vice-admiral, who suggested that the prince attend the demonstration of the ancient Ryukyu martial art of Karate as he was originally from Okinawa.

I was there that day in my capacity as instructor, having been ordered to take on this important duty and I have to say that I will remember the honor of performing that demonstration before his majesty to the end of my days. Later I heard what his majesty commented on during his trip to Okinawa. He was impressed by three things: first, the scenery, second the spritely water pouring out of the dragon spout at Shuri Castle and third Karate.

I had withdrawn from working as a school teacher just prior to this. The reason is, one day I was handed orders to transfer to a distant island and become the principal of the small elementary school there. This transfer was an honor as it was a testament to my ability, however my elderly mother had long been confined to bed due to asthma, and as an only child it was impossible for me to leave her.

Thus my thirty years as a teacher came to a close. However, I still wanted to make use of my knowledge and skills so, after consulting with the head of the Prefectural Library Majikina Shoko, (also known as Anko) and Sueyoshi Bakumonto, the editor of the Okinawa Times, I formed an association to support Okinawan students.

Following consultation with my Senpai, superiors, and friends, I decided it was necessary to form the Okinawan Shobukai, a group dedicated to the promotion of martial arts and martial spirit, in order to unify the teaching of Karatedo and I became its first director.

---

[49] Kanna Kenwa 漢那憲和 (1877~1950)

Funakoshi Gichin's son Gigo in Kiba Dachi

No Regrets From My Youth
Fighting a Typhoon

While these stories may be a bit embarrassing, I would like to write a little bit about my experiences when I was young and in the early years of my training.

First of all, there is the story of how we trained during a typhoon was recounted in a book by author Tokawa Yukio who was a member of our Dojo.

> The waves roared with disturbing intensity as rain began to pour from a darkened sky. The rain was so intense it, a person outside could hardly draw breath. Broken tree branches, sand and pebbles threw flew through the howling wind.
>
> Okinawans surround their houses with high stone walls and secure the roof tiles with mortar. But the winds are so tremendous sometimes attaining a velocity of 180 kph/ 112 mph, so that each house, no, in fact the whole town of Shuri was shaking and crying with fear. Everyone was huddled inside their homes trying to remain calm and praying each second that the storm would abate.
>
> However, to say that everyone was shut firmly inside their house would be a mistake. There was one young man living on the outskirts of Yamakawa Town who was determinedly battling the typhoon from the roof of his house.
>
> "A crazy person!" would have been what anyone who saw him thought. The man youth was wearing only a Shita-obi, loin cloth. He stood with his feet set wide on the slippery, uneven rooftiles with every muscle in his body tensed. In his hands he led a piece of Tatami mat straight into the oncoming wind. Clearly he had tumbled off the roof a number of times as his sturdy form was smeared with mud not just on his shoulders, but on his back, hands and feet. That mud combined with sweat was now being washed off in the pouring rain.
>
> He was a young man, maybe about twenty years old. He was small, barely 5 Shaku, 5 feet/180 centimeters, tall but his frame was wide and his muscles bulged and looked hard as iron. His hair was styled like that of a Sumo wrestler, with a topknot tied so it extended forward. It was pierced with a small silver Kanzashi hairpin indicating that he was part of the Samurai class.

Edo era illustration of a Nio guardian

However it was the committed expression on his face that stood out, resolute against the raging wind and rain. His large eyes, his prominent brow, standing with the intensity of a Nio guardian deity. His mouth was pressed tightly together as if saying *Un!* and his body, filled to the brim with power was bright red with effort.

The young man had both his elbows pressed firmly against his sides, his chest out and back straight. With his hips low he struggled with all his might to hold the Tatami mat into the wind. The point to focus on was his lower body. He was standing with his legs wide, as if he were sitting astride a horse. If anyone who knows about Karate saw him, they would immediately recognize the stance as *Kiba-dachi*, the Horse Riding Stance. This stance is unique to Karate and the young man was now standing outside at night during a storm, training with all his might. The wind continued unceasingly and if the young man had faltered in the slightest he would have gone tumbling.

Translator's Note: *The Man Who Chased Clouds*

One summer a typhoon. Okinawa is known as Typhoon Island after all as the storms always follow that route. That time Itosu Sensei gave Nenchin Sensei [sic. Probably referring to Gichin Sensei] a Tatami mat and told him to get on the roof and face the typhoon. In Karate there is a stance called Naihanchi, written with the Kanji Iron 鉄 Horse 馬. It looks exactly like a person astride a horse and the meaning of the stance is to remain unmoving no matter how much force is applied to you, just like a knight in armor on an armored horse. He faced the storm in that stance.

He made him do that?

He did. Of course, the roof of a house is hard to stand on, especially if you are trying to hold a Tatami mat into the wind. The wind was blowing something like a eighty or a hundred miles an hour, and one gust caught him and sent him tumbling off the roof.

As you would expect!

After that Itosu Sensei told him *Mada Shugyo Tarinai*, you need more physical and mental training. Then Itosu Sensei went out to show him how it was done. He went up and stood on the roof like a bronze statue, absolutely immobile. Nenchin Sensei, who at one point had held his head high in pride of his abilities, was forced to admit, *Alas, I must shamefully admit my Sensei is at a level that far exceeds what I ever hope to attain.*

Did Itosu Sensei ever make you do something severe like that?

He never made me do anything like that. However, from that day on, whenever there was a storm he would go up on the roof and face the wind. He continued doing this until he was in his middle years. In short, training was hard and painful.

The Man Who Chased Clouds 雲を追う男
By Togawa Yukio (1912~ 2004) 戸川幸夫
1960

Translator's Note: *Kiba Dachi*

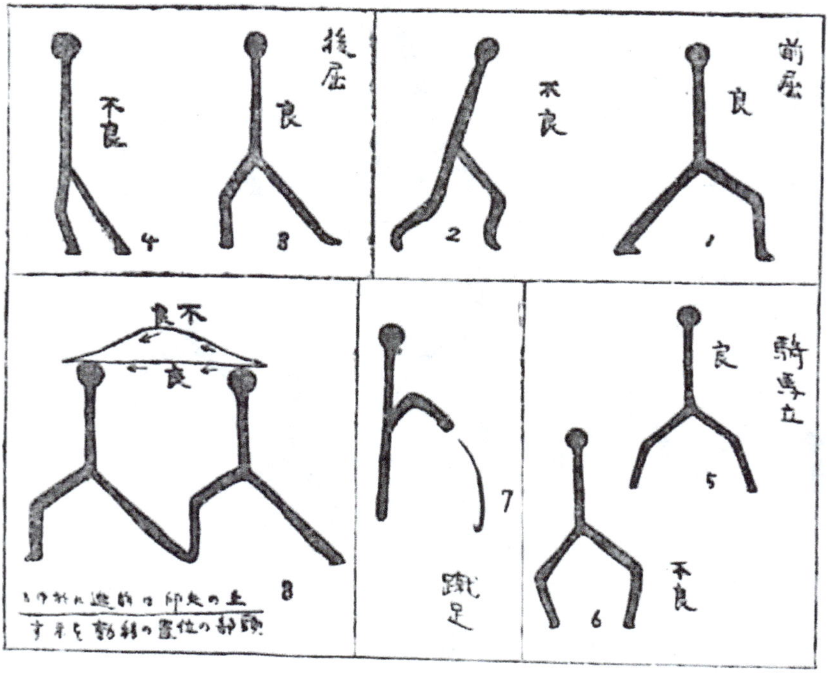

Karate Kenkyu 空手研究 *Karate Research* Volume 1
1934

Gokutsu | Zenkutsu

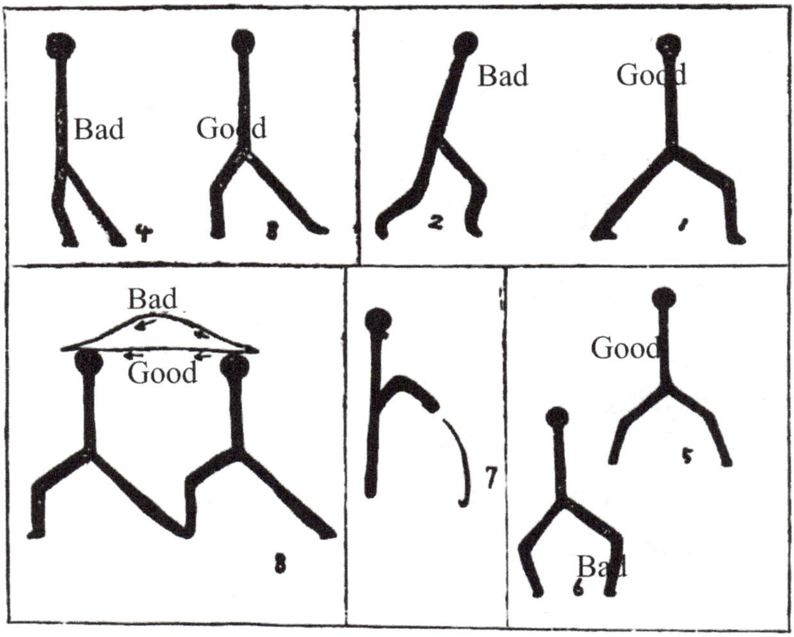

The upper arrow shows the incorrect way to advance. Your head should stay at the same level when advancing.

Kick | Kibadachi

US Army photograph (1950) of a Habu being milked for its poison. An antidote derived from Habu venom was developed in 1969

What I Was Taught By A Viper

No doubt most of my readers are already aware of the poisonous viper called a Habu that is native to Okinawa. These snakes are extremely common. While there is an small type of Habu called a silver viper, the average Habu grows to 3 Shaku (3 feet/90 centimeters,) with larger specimens growing to 2 Ken, 12 feet/9 meters.

The poison these snakes possess is quite strong, if you are bitten it is said your only recourse is to cut off that part of the body. If it's your hand then cut of your hand, if it's your leg, then sever your leg off. Failing to do this will mean the poison will travel throughout your body. Nowadays there is a fantastic anti-venom that can be injected so there is no need to go to extreme measures if you are bitten, however at the time Okinawans were quite afraid of Habu.

This is a story that happened one night.

There is a place called Sakashita that is half-way between Naha and Shuri and right at the top of that road there was a shrine to Kannon, the Goddess of Mercy. I was walking with my eldest son, who was in the second or third grade in elementary school at the time. I had given him the Chochin paper lantern to carry as we took the usual route to Asato Sensei's house. That part of the prefectural road was pitch black at night, but presently we came across what looked like two piles of horse manure stacked one upon the other. I though, *Ah, horse manure*, and continued on normally, however I soon discovered that it was a living thing.

My son was oblivious to this fact, but I pulled him to a stop. My son suddenly let out a yelp when he saw, from the dark mass on the road a pair of leering eyes and red tongue flitting out like a tiny flame. Startled, my son raced behind me.

With my son protected behind me, I took hold of the lantern in my right hand. Then I swung the lantern left and right in time with the serpent's breathing. It goes without saying that I kept my eyes on the snake the entire time.

I don't know how long this went on for but at some point the snake, with its eyes still locked on mine began to slowly slither sideways off the road and into the nearby potato field, where it disappeared. The snake had an impressively thick body and was incredibly long.

While I have frequently encountered Habu, this was the first time I have ever seen one coiled and ready to strike. I was still shocked how easy it was to make the snake retreat. So with a feeling of, *What was that all about?* I went over to the side of the road and shone my lantern over the tangled potato field to see if the snake was truly gone, when suddenly I was on guard.

The Habu had predicted that I would follow it and now the snake was poised, with its leering eyes fixed on me. In the end the snake slithered off. However I learned a powerful lesson from this incident.

People often talk about the persistent nature of Habu vipers, however what I learned had nothing to do with that. It was something else entirely. As we continued down the road I talked about this with my son.

It seems like Habu vipers know Karate. When the snake first slid sideways off the road into the potato field, it wasn't escaping. It was readying itself for the next attack. Being prepared to respond in any direction, in front, behind, right left, up, or down is the essence of Karate. This skill is something the Habu naturally possesses.

One Moment, Leaping Over the Sea

This next story, which happened when I was setting out for Tokyo for the first time, is a bit embarrassing.

Passengers on a Tenma boat in Naha Harbor 1927

The port of Naha, which is the most important one in our prefecture, is unfortunately too shallow to allow large ships to tie off to the cliffs. Thus large ships had to anchor out in the middle of the harbor, and passengers had to be ferried to land in Japanese style Tenma Sen, Heavenly Horse boats.

A storm had just happened to roll in that day and the rain and wind was intense. The crew helped us passengers into the small boats and eventually, after much bobbing up and down on waves, we made it to the boarding ramp of the ship at anchor.

Finally, it was my turn to board the big ship. The waves had gotten quite big. I had two large suitcases on the small boat. After waving goodbye to my family and Asato Sensei, who had come to see me off, I turned to face the big ship.

When the waves were calm, the side of the small Japanese-style boat was nearly even with the gangway of the main ship, but just as it was my turn to cross over, a large wave rolled in.

The little Japanese style boat began moving up and down dramatically and I wasn't able to step over onto the gangway of the large boat. Plus, I had two heavy suitcases in my hands.

So, I waited for a lull between waves and then seized that chance to plant my foot on the gangway of the big ship.

The moment I stepped, another huge wave rolled underneath, and the Japanese style boat began rising up and down. So there I was with one foot on the gangway, one still in the little boat.

Despite being born in Okinawa and raised in the town around Shuri Castle, I couldn't swim, in fact if you dropped me in the water I would sink like a hammer.[50] If I fell, that was it. The crew was shouting something at me, but I couldn't hear them.

Realizing I had made a blunder, I immediately shifted the heavier bag from my left hand to my right and then swung the heavy bag up towards the gangway. This pulled my body and remaining leg out of the tiny boat and I was able to leap to the safety of the main boat.

I had been in quite a pinch and if I had hesitated, what would have happened? I probably would have dropped down into the sea. Had I not been rigorously practicing Karate every day, there is no way I would have been able to make use of the weight of my luggage to enable me to leap onto the gangway. Truly Karate is a blessing.

---

[50] "Swim like a hammer" is a common phrase for those that cannot swim.

Later, when I returned to Okinawa, I heard that everyone in my family, as well as Asato Sensei, had all seen what had happened and gasped as I leapt aboard.

I went to pay my respects to Asato Sensei and he commented, "Welcome back! It is good to see you and I heard you did well! By the way, we were frankly all in shock at that scene on the day of your departure. That you were able to leap across was truly a marvel to behold. Yes, fantastic indeed!"

I turned red at all the compliments.

I am sure this applies to other martial arts as well. There are tales of people who practice Judo suddenly falling and, due to their training coming away unscathed. Such people credit Judo training for their lack of injury. Your daily *Tanren,* forging your body, will end up being beneficial at unexpected times and is therefore essential.

I can't recall exactly when it was, but one of my students Togawa Yukio answered a reporter from *Weekly Sankei Magazine.* He said something along the lines of,

> There are many reasons that I am glad to have studied Karate, but one of them is an experience I had during the war. I was serving as an information officer in the army and the plane I was riding in had to make an emergency landing in the ocean. Many of the soldiers as well as other information officers who I was travelling with died in the crash, but I was miraculously saved. Thinking back on it, I must have unconsciously avoided danger at the moment of impact.

This natural reaction of the body matches my experiences as well.

203 崎山御殿（御茶屋御殿） 首里市崎山

Ochaya Odon, the Royal Teahouse
Ryukyu Tatemono琉球建築 The Buildings of Ryukyu
By Tabe Yasushi 田辺泰 著 著者 (1899~1982)

To Win by Losing (Part One)

Just to the south-west of Shuri Castle stood Ochaya Odon teahouse that served as a vacation villa for the former lord of the domain. The teahouse dates from the Nara era (710~794) and overlooks the Pacific Ocean.

It was called Ochaya Oden, the Royal Teahouse, as the former head of the Domain would vacation here with his wife and personal assistants, to recover from his tiring administrative duties. The villa was a little over a mile from Shuri along a road was paved with stone and lined on both sides by lovely Ryukyu pine trees.

In later years, the villa was opened to the public for both touring and relaxing in. One evening I went there with Itosu Sensei and five or six other members of the Dojo for a moon-viewing party. Everyone was on good terms with each other, and we talked about Karate, sang songs and discussed poetry, completely losing track of time. When we finally decided to call it a night, it was already quite late.

A student younger than me led the way, carrying a lantern. We chatted as we walked down the stone path. We reached the point where the stone path intersected with the prefectural road, when the person in front let out a shout of, *What's this?* And then doused his lantern. We all reacted like a spring that had just been released, leaping sideways onto the main road with Itosu Sensei in the center.

The light from the moon was diffuse, not enough to discern faces, much less feet. However, it was clear we were facing four or five assailants who seemed shouting something. The man at the front of our group was trying to explain something, but I couldn't catch what he was saying.

I looked up at Itosu Sensei's face, but he was focused intently on the words that were being exchanged up front. Then he said, "Turn your back to the moon! Your back to the moon!" Those two instructions were all he gave. We had all been waiting for Sensei's command. I was sure that we were going to get a chance to show him the result of the endless days of training our bodies and minds in Karate. However, other than telling us to turn our backs to the moon, he remined silent. We all fell into uncertainty.

I can't say how long this lasted.

Finally, Sensei spoke to me, "Young Funakoshi, they may know Itosu, no, they may have mistaken us for a different group. Why don't you go over there and talk with them, but be casual about it." [51] And with that I began walking forward calmly.

One of the strangers said, "Someone's coming, someone's coming!"

As I moved forward, I could feel atmosphere all around me grow tense. The men each had Tenugui cloth covering their faces up to their cheeks, so it was impossible to identify them. As instructed, I politely asked if perhaps this wasn't a case of mistaken identity and told the men that Itosu Sensei was part of our group and that the rest of us were his students.

One of the gang said, "Itosu? I've never heard of that guy!"

Another one, seeing that I was rather small-framed,[52] cried, "What? You look like a little kid, why are you trying to get involved?!" And with that, he grabbed me by my collar. His grip felt strong and as I was thinking this might get bad, I began lowering my hips as I began putting power into them. However, from behind I heard the sharp voice of Itosu Sensei saying,

Kamii!

In Okinawa there is a tradition of giving a child a *Yomei*, a childhood name. My name was *Kameju*, "one who lives the long life of a turtle." People that were particularly close to me would call me *Kamii* instead of *Kameju*.[53] Hearing this name, I stopped dead in my tracts and waited for the next command.

"No fighting! Keep your hands down. Listen to what they have to say."

Thus, in a manner even quieter than before I asked, "Do you have some sort of grudge against us or is it…something else?"

Before I had even finished speaking, several people who had apparently been drinking walked out from a small stone lined side road singing. When their path crossed ours, they immediately sensed the strange tension and shouted, "Hey a fight, oh my oh my, this is sure to be interesting!"

However, the next moment, the manner of the new group changed entirely. They all began straitening the collars and belts of their

---

[51] Itosu Sensei is referring to himself.

[52] Funakoshi Sensei describes himself as a *Kohyo* 小兵 small soldier, referring to someone of slight build.

[53] The Kanji 亀 meaning turtle is read as *Kame* in Japanese but *Kamii* (or *Kaamii*) in Ryukyu.

Kimono and one said, "Itosu Sensei I presume? Indeed, you are Itosu Sensei, what on earth is going on?"

He then turned to the group with towels covering their faces up to their cheeks. "What do you all think you're doing? This man is Bushi Itosu, the martial arts practitioner.[54] The others accompanying him are his students. Even if there were ten or twenty of you it wouldn't make a difference when going up against them. You need to apologize now, I mean right now, if you don't want this to go any further!"

The ruffians, who were at first startled by the interruption, and were muttering fiercely, gradually lapsed into silence. We too left the scene and headed home, though at Sensei's direction, we took a different route from the one we set out on. Itosu Sensei did not mention the encounter the whole way home, but when we took our leave from him he said one thing, "You all did a fine job tonight. It is important that *Shugyosha*, those on the path of Karate training, behave in such a manner. Be sure not to talk to anyone outside the Dojo about what happened tonight."

Later I heard that the group of ruffians we encountered went to Itosu Sensei's house and politely apologized for their actions. It turned out that the men were *Sanka*, semi-nomadic people. They lived in a village that made Ryukyu Awamori, a strong traditional alcohol,[55] and considered themselves to be something like the Sake Gura Otoko, Sake brewery toughs, you see about these days. They were quite proud of their strength and simply saw us a way to test themselves.

Looking back it was clear that Sensei warning us not to raise our voices and, after that, avoiding further complications by taking a different route home, I was able to realize the extent to which a *Bujutsuka,* true martial artist, carefully evaluates each situation. If Itosu Sensei had not called out *Kamii!* I would have blundered into a

---

[54] Bushi refers to a Samurai on mainland Japan but a martial arts expert in Ryukyu/Okinawa.

[55] *Awamori* is the oldest distilled alcoholic drink in Japan, and is believed to be the predecessor of *shochu* (another distilled Japanese spirit). The technique for making *awamori* is said to have remained unchanged for over 500 years. Indica Thai rice is soaked in water and sprinkled with black koji mold, which is then fermented with water and yeast, before undergoing a single distillation to complete the process.

-Official Okinawa Travel Guide

huge fight. Thinking back on it now, I realize how green I was at the time and feel quite embarrassed.

To Win by Losing (Part Two)

Following this incident I was complimented by both Itosu Sensei and Asato Sensei.

I must say a word about my wife's family. For many years they had been working to improve the Satsuma sweet potato plant, and had found some success. However, following the Meiji Restoration in 1868, their fortunes had turned and they moved to the village of Mawashi, about a mile down the road from Naha. While my wife's father was a member of the Obstinate Party, he lived simply and took things as they came. He was a bit of an unusual fellow, following the philosophy of *When the weather was good, he farmed, and when it rained, he stayed home and read books*.[56]

I am pretty sure this happened during the Obon or some other festival. My wife and children had gone to pay their respects, but it was getting quite late so I decided to go myself, light some incense as an offering and then walk home with them.

So, I walked between the flooded rice fields, through plots of millet and down the path through the pines trees, a path that, even in daylight, frightened young girls. As I was passing through a section of woods, two men leapt out from where they had been hidden in the shadow of a pine tree and moved smoothly in front of me. I took one or two steps back and judged at a glance that they were looking to steal whatever possessions I might have.

"Hey! Don't just stand there mute, say something! Or maybe you understand what's going on without having to talk?"

I didn't respond, just shifted my old, ripped lantern, in front of my right hand, as a first line of defense. As they had Tenugi cloths over their faces, no matter how hard I tried, it was impossible to see what they really looked like.

"Well, why don't you say something? Say, 'Good evening' or 'It sure is hot today! Can't you offer some kind of greeting? You listening little fella? If you give us a lot of trouble it will only hurt more. Hey! Come on, spit it out."

---

[56] *Sei-Ko-U-Doku* 晴耕雨読 If clear farm, if rainy read. Separating yourself from the hustle and bustle of life.

As the man continued to speak, he got more and more irritated, while I became calmer and quieter. Glancing at how one of the men had his hand in a fist I realized he was not a *Shugyosha*, practitioner of martial arts. I also had time to evaluate the other man who was holding the thick broken end of a Bo staff. I spoke to them in a calm voice, "I think you have mistaken me for someone else. I have no idea what is going on, but how about if we talk this over."

"What? You little punk, better if you keep quiet, who do you think we are? Stop trying to play games!"

The two men began inching closer, but quite surprisingly I was comically calm, almost beyond calm. I said, "It appears that this is going to devolve into a fight, however, be aware that this will not go well for the two of you, so I recommend you desist…"

At this the two men grew furious and tensed their bodies, obviously readying themselves to leap at me. Before they could I cautioned them again,

"Wait a moment. The reason this will not go well for you is because it is clear you will win. If you know from before the fight that you will defeat your opponent, doesn't that mean you are losing out?"

Hearing this, the two men suddenly grinned showing their white teeth. "Well, looks like this young fellow has given up! No point in trying to beat up a punk like you. Fine, hand over your cigarettes or whatever else you got, come on, make it snappy!"

However, I have never been a person who enjoys tobacco, further as we were a poor household I had no money on me. Since I felt I should offer up something I pulled out the package of Manju cakes that I had bought. I had been planning to place them as an offering on the household shrine at my in-laws house.

"This is it?! Well, suppose it's better than nothing. Hey, be careful around here, this is Habu territory (basically stating that the area was dangerous.)[57] Be careful."

And with that, the two men disappeared in the darkness.

A few days later I happen to be with both Asato Sensei and Itosu Sensei, and I happened to mention this incident when a related topic came up. Upon hearing my tale they both clapped their hands together in praise.

---

[57] Brackets are by Funakoshi Sensei

Itosu Sensei spoke first, "Well, you certainly handled that situation well! It seems that teaching you Karate has borne fruit, you did good and should be praised!"

Following Itosu Sensei's kind words, Asato Sensei spoke, with a grin on his face, "So, since you no longer had the Manju cakes, what did you place on the shrine at your wife's parent's house?"

I have to admit to being a bit shocked, as my wife, children, elderly parents and other relatives were right there and I hadn't mentioned this aspect to them.

"Since I had nothing," I replied, "I bowed my head, lit some incense and offered a heartfelt prayer."

"Ah, well done, well done," declared Asato Sensei. "Little by little you seem to be developing a sense of what Karate is."

My face had turned red at all this praise. Typically neither of my Sensei would offer words of praise during Kata training, so to now be given a passing score, I was somewhat overwhelmed.[58]

---

[58] Itosu Sensei and Asato Sensei gave him *Kyudaiten* 及等点 which refers to the lowest possible passing score in a test. In short, "You passed but just barely so keep at it."

Outer gate and main temple of Sogenji Temple, which was burned down in the Second World War

The Outcome of an Arm Wrestling Match

I believe this story is from sometime after I had turned thirty years old.

One night, I was walking from Naha back to Shuri. After passing Sogenji temple the road became very lonely indeed. There was a graveyard on the left side of the road and nearby a large pond that porters used to wash and water their horses. Just where the road began to curve to the right, there was a grassy spot some six Tatami mats in size.[59]

There was a big tradition of Okinawan youths testing their strength against each other, and just then there was a group of young men arm wrestling each other.

As I mentioned earlier, *Udezumo*, arm wrestling, in Okinawa is done somewhat differently from the mainland. First there is a stone about 2 Shaku, 2 feet, across and 3 Shaku, 3 feet, high is set up. It sort of serves the same purpose as a *Dohyo*, the mound Sumo wrestlers compete on. Two opponents stand on either side of this stone. Every village will have one of these stone slabs.

I was a big fan of arm wrestling. Further, I was pretty confident in my abilities. Thus, I stopped to watch, and after a time someone called out to me.

"Hey you! Just standing there watching while other people do arm wrestling is rude. If you like it that's fine, if you think you have a chance to win, then come over here and give it a try."

I figured if I hung around too long it might turn troublesome, so I replied

"Thanks, but I am not looking for a match."

Having given my answer I started to walk off. However, I heard the rustling of two or three fellows coming after me.

"Come on back, come on back! You stood there watching silently and now suddenly you say you don't want a match and are just going to go home? That is impolite and we can't allow it!"

At that point they had already grabbed the sleeve of my Kimono and were clearly set on dragging me over to the arm wrestling spot. I relented and walked over to where an older fellow was seated in front of the stone with his back to a sugar cane field. He seemed to be the strongest person there, not the referee as I had assumed.

---

[59] Six Tsubo 坪 or 213 square feet.

Translator's Note: Okinawan Arm Wrestling

Two brothers doing Ude-Oshi, arm pushing, under the shadow of the fountain palm trees as their father watches. In this type of arm wrestling combatants do not grip palms, rather they put their wrists together.

There were three types of arm wrestling, *Udei kakii* 腕掛 arm-hooking, *Nubi kakii* 延掛 extended-arm, and *Iishi ude* 石腕 stone-arm. Those who practiced Karate tended to prefer *Udei kakii.* This is because in *Udei kakii* you don't place your elbow on a platform or otherwise support it when dueling. Thus, the person who could throw off the other person won. Also, *Nubi kakii* is done with the arm kept extended, without bending the elbow. *Iishi ude* is the same as ordinary arm wrestling, but there were methods such as placing the elbow on a Chikara Ishi, test of strength stone, or holding a pole, like the ones used to carry goods, in the left hand and resting your elbow on that. In any case, since it was an age with almost no entertainment, what the young people did was extremely simple.

-*Discussion of Modern Karate History* 近代空手道の歴史を語る
Gima Shinkin, and Fujiwara Ryozo 儀間真謹・藤原稜三
1986

Having come this far it was too late to try and back out so I relented and took on the weakest competitor and I greeted him with a *Onegaishimasu*, If you please! But my opponent only glared at me intensely. As it turned out I was able to defeat him easily.

After that, a different man took his place. I defeated him as well. Then a new man took his place. I won that bout as well. Three victories in a row.

"I doubt I will be able to beat the next one?" I surmised.

Despite my dire prediction, I won that match as well. Though I quite enjoyed this type of bout, and was curious how far my strength could take me, I probably should have retired and headed home. At that point I had beat five or six men in a row.

Finally, there was only one other person left besides the older man who looked like a referee. The older man called the remaining combatant by name, saying,

"It's your turn next."

Now there was a great deal of tension in the air of a sort that was hard to describe.

"Sorry, I am afraid I've had enough. Further, it seems highly unlikely that I will be able to win against you. Please excuse my departure…" I said while rising. However the man selected to be next continued to stare fixedly at me with his arm out and ready. His whole manner conveyed, do you want to challenge me or not? So, I gave in, and the match began. As it turned out, I won.

"Thank you very much, but now I must leave."

I set out and walked back to the road and started heading for Shuri again. However, as I walked I began to sense the same strange feeling of tension that I felt before I took on my last opponent. The feeling grew and built, and I realized that I was being followed.

As I suspected, after walking another mile I felt the presence of people behind me.

When I had set out that day, it looked like rain, so I had a western style umbrella with me that I was now using as a walking stick. To protect my head, I opened it part-way and placed it over my head.

Suddenly, the atmosphere around me seemed to grow thin, and that was when it happened. The first hook landed. I was able to block it without changing my stride. Judging from the *Keihai*, sense of what was around me, there were more than one or two of them behind me. Just as I was realizing this, a second attack came. The second man had a fast, accurate punch. However, I was able to deftly direct it away.

The third strike was a kick aimed at my legs. I responded by shifting my right leg out in a flash and, as the man had already committed to his strike, he was caught off guard. Since I had dodged, the third attacker completely lost his balance and fell in a heap directly in front of me.

Just then I heard a voice from a person at the back of the group.

"Who is this guy? He seems to have a good knowledge of some martial art." Then to another man he said, "You there, go get him!"

There was no mistaking that voice, it was the man who had been judging the arm wrestling contest. The man he set on me was my last arm wrestling opponent.

The man didn't say a word, but just built in intensity as he closed on me before letting his fists fly. I deflected all his strikes. Then the same voice shouted,

"Try this on for size!"

A man had closed on me using *Shinobi Ashi*, silent walking, and, when in range, swung at me with some sort of club. However, the moment he made his attack, I leapt back. His club just clipped the edge of my sleeve.

At this point I still had the Western umbrella covering my head, but glancing around I saw there were seven or eight men in a tight clump in front of me. It didn't seem as if anyone was going to try and take another wild swing at me, so I started walking off. I even started signing a poem in a quiet voice. Though I was primed, I didn't hear the sounds of haggard breathing or stomping feet in pursuit.

Frankly, I found the whole situation rather mysterious, but at the same time I regretted my actions somewhat.

"If I had been facing ten opponents, would I have still been able to walk away humming a poem? What if there were twenty men? Or thirty? Or more? It seem clear to that I have become a bit overconfident in my training."

I get an uncomfortable feeling whenever I recall this incident.

Translator's Note: Bengatake Stone Gate

188 弁ヶ嶽石門(昭和の初め)。「下り口説」に「あれあれ拝めお城元、弁のお嶽も打続く」とあるように昔から弁ヶ嶽は船舶の目標ともなっていた。

An early Showa Era (1926~1939) photo of Bengatake Stone Gate 弁ヶ嶽石門 The entire peak of the mountain is considered to be the sacred body of the spirit of the mountain. There was a poem by sailors from Satsuma Domain that went,

Look there! It is the Bengatake shrine near Shuri Castle! I can see it clear as day!

Mistaken for a Robber

My grandfather's house was in the Teracho Town section of Naha, not far from a sacred forest known as Bengatake. For people that travel to Okinawa by boat, this is the first thing they see. The reason the forest is sacred is because it surrounds a shrine to the god of wealth and good fortune, which is why people, not only from Shuri but from other town and even distant Naha, travel great distances to pay their respects.

The forest looks down on Nakagusuku Bay and there is a road that slopes down gently towards the port towns of Nishihara and Yonahara.

This happened to me one night. I was walking up the gravel road from Nishihara village towards the summit of Bengatake. I was just nearing the highest point of the mountain, when I realized there was a man running straight at me so fast he appeared to be tumbling down the road.

Though I was just at the point of cresting the ridge, I shifted my body sideways in a flash so I was perpendicular to the man who was charging directly at me as if he wanted to crash into me and take us both to the ground. Though it was quite dark, I could tell by the way he held his body and the way he moved that he was not a *Ken no Tsukaite*, practitioner of Kenpo. I registered all this information in an instant. However, I had no time to call out to him before the man slammed into me.

Having no other choice, I deflected him with *Uke no Kamae* and the man flew sideways head over heels. Beside the road was a millet field but when I looked, I could find no sign of the man.

"I wonder what happened?" I thought. "I doubt he could have run off in such a short span of time…Or maybe he struck his head on a rock when he fell and got knocked unconscious."

Thinking that was a possibility I began searching through the field, shifting aside stalks of millet, but the only sound was of leaves brushing against each other. No so much as a moan.

Thinking this was a bit mysterious I walked on about 2 or 3 Ken, 12~18 feet, until I came to a small depression, where a black watermelon looking shape was moving. As it turned out the man had the misfortune to end up getting thrown into a farmer's cesspool due to me redirecting his powerful charge.

Feeling bad for the man, I extended my hand out to help him despite the powerful stench. However, once out of the pit, the man didn't offer a word of thanks but once again began charging pell-mell down the road.

I was so surprised that I just stood staring after the fleeing man. It must have seemed like a fox had taken over my body. Just then from the opposite direction came the sound of a police whistle,

Piiii, Piiii

Suddenly two or three dark shapes came running up.

"Oh, it's the police." I thought.

I was just stepping out of the way to let them pass when the three officers raced up and grabbed me.

The three policemen held me firmly and turned me around so my back was facing the top of the mountain and they readied to haul me back down the road. Siezing this moment I shouted,

"Let me explain! I am not the man you are looking for!"

However the three men paid me not the slightest heed, with one saying, "Listen, we are the police. Now come along quietly!"

Even as he was saying this, another of his fellow officers was readying to tie me up with a *Hojo*, police rope.

I thought to myself, "Wait a minute, they are probably after the guy I threw into that cesspool!"

Realizing this I said, "Wait, I believe you have the wrong person!"

A policeman replied, "You liar, even now can't you just admit you are guilty?" Clearly he had already decided I was the criminal they were looking for.

I stood in a right Kamae and said, "Just now a man slammed into me. I knocked him aside and he fell smack into the cesspool. I think you all are confusing that man for me!"

While at first they didn't seem to believe me, the three officers put their heads together and began to talk it over. Satisfied they admitted that they had made a mistake and apologized for their poor treatment of me. Then they began peppering me with questions. What did he look like, how old? And so on.

However, as the man had slammed into me out of the pitch black of night and when I swept him aside he came up covered in *Kogane no Mizu*, golden water, I had no idea how old he was or what he looked like.

Obviously, the direction the man had escaped was clear, we all began descending the hill in a line, until, from below us we heard the sounds of rustling and voices. It became clear that they were also police. The police officers had pursued the suspect from two directions.

Soon, a voice from below called, "Hey! There is a man collapsed on the ground here!"

The police and I all ran to the scene. We had to run for two or three Cho, about a quarter mile, to the site. It turned out the man I had encountered previously was laid flat out in a potato field. Someone lit a match and by the light determined he was unconscious. It was clear by the foul smell that his clothes were soaked with liquid from the cesspool.

The police determined right away that he was the man they were looking for. They immediately began tying him up. However, as he was unable to resist, it seemed rather cruel to tie him up and one of the officers suggested, "First of all let's give him a bath. He can't really resist and we certainly can't haul him into the station in this condition."

So, as the man was beginning to come around, they loosened the ties they had on him. The whole group ended up walking right past where I live on the way to the well to wash the guy off. When I saw his body I was surprised.

He had blood on his right side and a bruise on his left hip. The first was probably from when he fell into the cesspool, but the other one seems to have been where I instinctively kicked him as he tried to slam into me. I probably overdid it there.

I was feeling a bit bad for the fellow when I heard what the police were saying. They said that not only was he an armed robber but had also assaulted women, stolen, burgled and committed innumerable other crimes which had terrorized the local village of late. Several police forces had joined together and had been chasing the man over the past two or three months. The moment I heard that, my sympathy for the man dropped significantly.

Translator's Note: Hayanawa, Fast Tie

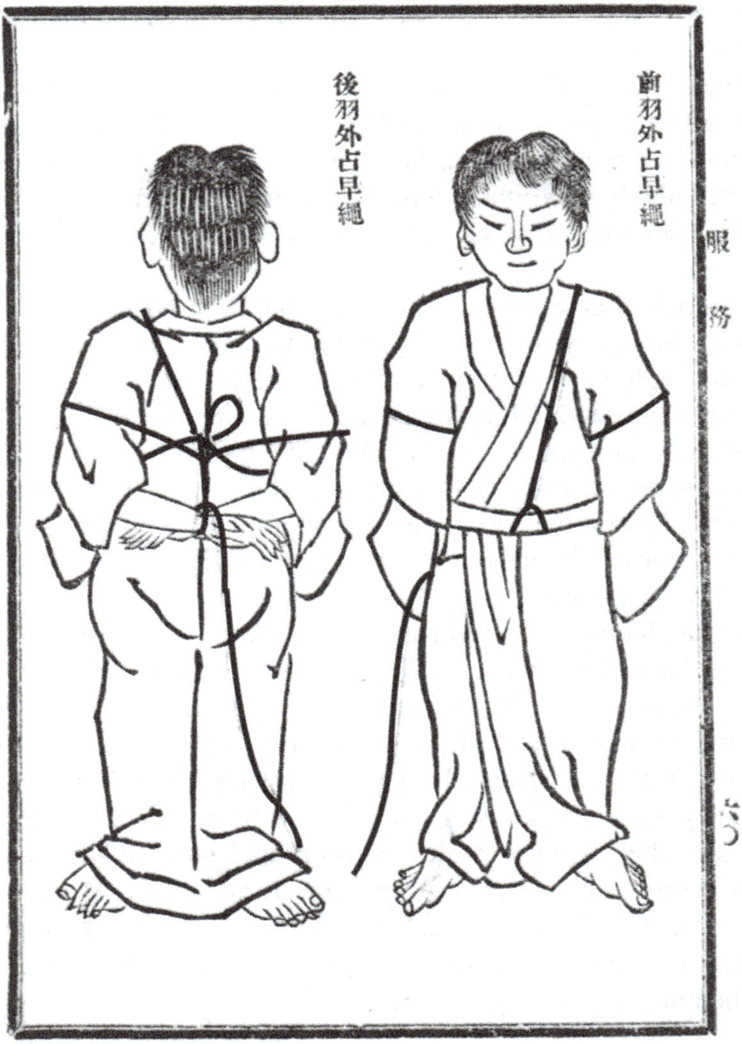

Up until now there has been no official way to tie a suspect. Police officers are expected to train these techniques carefully in order to carry out their duty. Each police department as well as each officer should be following these regulations. Illustration showing how to do

Hayanawa, Fast Tie on a Suspect
How to Use the Arresting Rope Hojo 1888

Being Assigned a Mediator Leads to A Promotion

I think I will write about one more episode.

Along the northern part of the main island of Okinawa is a beach called Kokuryo, which is where Motobu Village is located. Further along that beach from Motobu there is the remote hamlet of Jahana.

Following the Meiji Restoration, Jahana became a refuge for *Fuhei Shizoku*, Discontented Former Samurai from Naha, so it was an area where you could expect to find trouble. If there was unrest or disputes in the Kokuryo region you could be sure that it originated in the small collection of houses known as Jahana.

This happened in the middle of my carrier as a teacher.

The strife in the area was centered on the small village of Jahana, but the conflict drew in people from the surrounding villages. Despite the fact that the local magistrates had brought in police from the surrounding jurisdictions, the conflict could not be resolved. In the end, a request was put in for additional officers from police headquarters in Naha.

This was more than a simple matter of detectives investigating an incident, the group heading to the area included both village and regional assembly members (at the time such assembly members existed) who grew up in that area as well as Okinawa prefectural assembly members and officers from the Okinawa prefectural government.

As it turned out, for some reason, I was requested to accompany the group.

Since I was just a teacher, I found it difficult to believe that I would be of much use, however having been invited I got permission from the principal to take a leave of absence.

Of course, at that time there were no automobiles around so we climbed into three or four horse drawn carriages and set off at a brisk pace for Jahana. We travelled at a quick pace so obviously, as the horses clearly can't run at that pace the entire 20 Ri, 49 miles, we had to change horses twice along the way. This meant that we didn't arrive at Jahana until nightfall.

Once there, I wasn't entirely sure what to expect, after all Jahana was a village full of members of the Samurai class, and I didn't know what we were going to do if they decided to get rough. And it goes without saying that there would be more than a few men amongst them who were good with their fists.

Thinking back on it, the reason they brought me along was probably to assist if negotiations didn't go as planned, but at the time it hardly seemed as if the situation was going to be resolved. The atmosphere in the village was tense and hard to bear.

However, suddenly one man from each of the opposing groups stepped out and began walking towards me. Looking at them closely, I realized that I knew each of them.

The two men, each from one side of the feuding groups, greeted me in unison.

"Mr. Funakoshi, we certainly did not expect to see you here!"

I replied with, "I have to say, I am not entirely sure why I was invited to be part of this group."

One member of the negotiation team commented, "Seeing as how you seem to be on friendly terms with both sides, why don't you see if you can't serve as an intermediary."

So, with the agreement of both sides, and without any of the other negotiators, I was now responsible for resolving the dispute between these two parties.

Obviously if things went well, the situation will be resolved amicably, however failing to do so meant being at the mercy of one group or the other. I could only image how that would turn out.

So, with great trepidation I began the negotiations with the two parties. I thought to myself, "If I make a mess of this…someone from one side or the other will try and punch me in the eye! Frankly, if it was just me that ended up taking a thrashing that would be fine, but this situation seems liable to spin out of control and get big enough that its effects could be widespread. This goes far beyond trying to solve an everyday problem."

My approach was, above all, to remain polite and sincere to both sides while listening to what they had to say with an open and calm mind, without preconceived notions. It goes without saying that more than once I felt the fury in the onlookers and was sure a member of one group or the other was on the verge of leaping straight at me and trying to hit me.

However, if I dropped into a Kamae at that point, it would only encourage violence while, at the same time, shattering any authority I had as a negotiator. Thus, I cloaked myself in calmness as I faced enraged members of the community. I allowed each side to say what they wanted and didn't interrupt them until they were done.

Finally,...after two days I succeeded at finding common ground between the two while, at the same time, allowing both sides to keep their honor intact. Having concluded the negotiations, all the tension I felt in my body was swept clear. The "hand clapping ceremony" to signify the conditions were agreed to was held at what is now Jahana Elementary School.

About a month after this I was called to the prefectural board of education where I found a long-awaited transfer to an elementary school in Naha was waiting for me. My transfer being approved was probably also due to my work dealing with the children wearing topknots that I mentioned previously.

At that time I was teaching at an elementary school in Ginowan, Nakagami Region and wanted to transfer to a more central school in order to be closer to Karate training.

In other words, it was thanks to Karate that I was able to receive a long-desired transfer which enabled me to further my Karate training. I would like to add that while the negotiations took quite a long time, the experience was well worth it, and any Karate practitioner would benefit from such an experience.

The Compassion of an Old Samurai

This story is from when I was still an assistant teacher.

I had just been transferred to an elementary school in Naha and my wife was still saying at her parent's house in Shuri, so I had to walk just under 1 Ri, two and a half miles, to school and back every day.

That day there was a teachers' meeting and it ran quite late. On top of that, by the time the meeting finished it had begun to drizzle rain so I hailed a Rickshaw.

Just on the border between Naha and Mawashi, there was a temple called Sogenji Temple. The road is rather dull in this area, so passengers and Rickshaw drivers often begin chatting at this point to pass the time.

I struck up a conversation with my driver and, though I couldn't see him, as the rain-hood was pulled over the seat, from his way of speaking and accent I could tell that he was from Shuri and noted that he used *Shizoku Kotoba*, Samurai phrasing, in his speech. Further, I was sure I recognized his voice.

Rickshaw circa 1868

While I am hardly a person who speaks in a coarse manner, the way the man pulling my Rickshaw spoke was very polite, despite the fact that he only responded briefly to my questions.

"This is certainly very curious!" I thought.

Since my interest was now peaked, I addressed him directly, "Driver, sorry, I need to relive myself..."

With that the man stopped. Exiting the Rickshaw, I noted that while he was a bit on the tall side, he was as thin as a crane. He was wearing a Manju hat pulled down at an angle, as if he were trying to hide his face as he waited for me to step down out of the Rickshaw.

However, seeing how he held the *Kajibo,* handles of the cart, I could tell he was no ordinary Rickshaw driver. The moment I stepped out of the conveyance I tried to catch a glimpse of his face but he quickly shifted his body and I wasn't able identify him.

At this point the rain had begun to taper off and a hazy moon was struggling to emerge from the clouds. As I live along the prefectural road, I stepped down into a depression before a field of millet to do my business, and tried to catch a glimpse of his face from that vantage, however he never raised his face. Having watched him for some time, I could tell by the way he held his body that he was a man of some martial skill.

Finally, I came up with a plan.

I said to the driver, "I have been riding for a while now, and am ready to take a break and stretch my legs. Let's walk together to the top of the incline."

So the Rickshaw driver and I walked side by side up the slope. I was sure at some point the man's guard would slip and I would be able to get a look at his face.

The slope in question was a tough one for drivers to climb. During the day some of them even went so far as to push the cart from behind.

As we walked I tried to engage him in small talk. However the driver wouldn't walk beside me and seemed to always find a way to remain two or three paces behind me.

Eventually we arrived at Heitai Oka, Army Hill, where the road began to bend. At that moment I quickly turned around and, as I grabbed the wooden rail of the Rickshaw, I tried to steal a glance under my driver's hat.

However, the man was quicker and used his right hand to pull his hat down even further. I noted that his movements were fast and nimble.

For a moment I had no words, "…"

Just then something occurred to me and I took off my hat and asked, while leaning over to look up at him, "Pardon me, but aren't you Mr. Suekichi?"

The man gasped and suddenly stopped moving. "I am sorry, I believe you have mistaken me for another." The man said, rejecting my guess.

So there I was, with one hand on the wooden rail and the Rickshaw driver standing there with his had still covering his face.

Suddenly, he took his hat off and dropped down onto one knee, bowing his head, leaving me holding the rail of the Rickshaw.

"So it is you!" I said.

All I could see of the man was his topknot, but I was absolutely certain it was Sueyoshi.

Translator's Note: Early Show Era Bojutsu

棒術組手　唐手が主として首里那覇などの如き都會地を中心に發達してゐるのに反し棒術は田舍の方が盛んである。棒は四百年來武具を撤廢されてゐた琉球に於ける唯一の武器である。

Bojutsu Kumite, Paired Staff Techniques
While Karate is spreading rapidly in Shuri, and other big cities local areas are seeing more and more people studying Bojutsu. The Bo was the only weapon available to Ryukyu people after the ban on weapons.

棒踊り　村の踊りの中で一番雄壮なものは棒踊りである。渦巻に行進したりほぐれたりする度毎に発するかけ声の勇ましきと団体的訓練の行き届いたことなど男性的舞踊の粋である。

Bodori, Staff Dance. The most impressive display of warrior spirit at village dance festivals is undoubtably the Bodori. Each time the group advances in a spiral and then unwinds, their spirited shouts, together with the precision of their well-coordinated group training, display the very essence of a masculine dance.

Rushing to take his hand I said, "I am Funakoshi. I was just trying to find the answer to a question I had, I apologize for my terrible rudeness in pressing you for an answer! However, I must ask, whatever caused you to become a man who taxis people about?"

I too had dropped down onto one knee.

This man was a high-class Samurai who served the former lord of the domain and a Karate Senpai of mine. He had, in his later years founded a school of Bojutsu (*Sueyoshi no Kon*, Staff Techniques of Sueyoshi.) Due to this past, Sueyoshi did everything he could to prevent himself from being recognized. He finally replied, but simply repeated,

"I am ashamed, please, I beseech you, do not tell anyone of this."

Clearly I was not about to get back in the conveyance. The two of us walked the rest of the way to Shuri with the Rickshaw between us, talking of Karatedo, Bojutsu and many other things. The man could only be described as *Seihin*, poor but honest, and his wife had become ill. Now he was leading a tough life of working as a farmer in the daytime and switching to being a Rickshaw driver at night. (At that time in Okinawa there were two categories of Rickshaw driver, *Hiru-guruma*, "Morning Carts" and *Yoru-guruma* "Evening Carts," and the two groups had strictly defined hours of operation. Thus, former members of the Samurai class frequently chose to work as "Evening Cart" pullers. Therefore people were advised to mind their language when addressing a Rickshaw driver in the evening.)

If Sueyoshi were looking to find success, then there were many roads available to him. However that kind of work did not agree with him. It may have been the pride of a Samurai, or maybe integrity, but I was inspired by this tragic but stalwart fellow.

I often reflect on how expertly the man stood with his hand on the *Kajibo,* both his grip and the way he held his body were something even a trained practitioner would be hard pressed to imitate.

Unfortunately, shortly after I moved to the imperial capital, he passed away. However, whenever I think of that episode I can't help but reflect that he represented the epitome of the traditional Samurai.

Female rope is 25 Ken, 150 feet long "Unchanged since old times"

Male rope is 10 Ken, 60 feet

Yonaguni Hara O-Tsuna Hiki 与那原大綱曳 1983

The Ultimate Tug of War

The most anticipated event of year for adults is the Tsuna-hiki, tug of war, competition. It is always held during festivals or other big celebrations. However, I imagine the way it is done in Okinawa differs from other prefectures. It is far more thrilling and taps into the martial bravery of the people involved.

This may be due to influence from China, but the rope, if it can be called that, is thick enough to require two people to wrap their arms around it. Attached to it are numerous smaller ropes, called *Mezuna*, women's ropes, giving it the look of a centipede.

19th century illustration of Ryukyu tug of war

There are two of these ropes, each of which is as thick as a tree, and at one end of each a large loop is made. Before the competition can begin, a thick pole of oak is slid through the loops at the ends of both ropes. The moment this is done, both sides begin chanting, ringing handbells and pounding on drums. The judge, who stands between the *Tsuna-Gashira*, head man from each team, signals the start of the contest by stepping on the feet of each man. The judge then stands *Nio-dachi*, with his arms crossed like a guardian figure, watching the duel and shouting encouragement to both sides.

Note that inserting the wooden core through the loops is quite a task, and is a major part of the judge's job. If not done carefully you can get caught in between the oak pole and the ropes and get crushed to death. Thus this job is one that can be hectic. For that reason, there is a rigid set of protocols to follow when inserting the wooden core which can seem troublesome at the time, but due to the danger involved no one will allow these procedures to be altered or eliminated.

The groups cheering on each side were roaring like they were going into combat. They had flags out and were pounding on drums cheering on their side and naturally, from within that great confused hubbub a winner emerged. It was not uncommon for a great fight to break out after a winner was declared, with severe injuries and even deaths on both sides.

Thus, the judge had to be a person nominated by both sides, as he may suddenly have to act as a negotiator between the two parties if a brawl started.

Shuri was the town surrounding the castle where the king resided, so the competition there was much fiercer than in other towns and villages in Okinawa. However, when I was working as a teacher in Naha, it became the prefectural capital and that, combined with the expansion of the port facility, meant that Naha gained in prominence with the additional effect that the tug-of-war festival grew to impressive scale.

I was frequently honored to be selected as a judge for the Naha tournament and, fortunately, never had a fight mar the proceedings while I was in charge.

Something I realized when judging these contests is that after I signaled the start of the contest to the *Tsuna-gashira*, head men, and then stood with my arms crossed like a guardian king, is that if a side was committed to recklessly throwing everything they had into the contest without thought or strategy, that side, more often than not, couldn't go the distance and lost.

On the other hand, if a group had *Kyo-Shin-Tan-Kai*, an open mind with no preconceived notions, as the overarching philosophy, they would, somewhat to my surprise, be able harness their strength and apply it effectively.[60] This is nothing more than an observation from everyday life, but I think there is a connection between this and the philosophy of Karatedo and we can learn from it.

---

[60] *Kyo-Shin-Tan-Kai* 虚 心 坦 懐 with utmost candor, with no preconceived notions

Support of Master Artist and Calligrapher Kosugi Hoan

I believe this was towards the end of the tenth year of Taisho (1922.)

The Ministry of Education held the first *Kobujutsu Tai-iku*, Old Martial Arts and Physical Fitness Exhibition, at what was formerly the Girls Upper Level Teachers School in Ochanomizu.

Okinawa Prefecture would also be participating in this exhibition and someone from the prefectural board of education contacted me and asked if I would be interested in introducing the martial art unique to Okinawa in mainland Japan. This was right around the time I had quit teaching and realized this was an incredible chance to introduce Karate to the mainland. I immediately agreed to go to the capital city.[61]

However, Karate is a unique art and planning out how to explain to spectators, who had never heard of these techniques before, what they were about to see was a difficult task. So, after a lot of thinking, I decided to take pictures of Karate techniques, how the hands and feet were used as well as Kata and Kumite and arrange them on two scrolls.

The exhibition was a big success and Okinawa benefited greatly from going to the trouble of sending a representative. Further, I was delighted that Karate had made inroads into mainland Japan, a solid first step.

Originally, I was planning to return directly after the event ended, however during the proceedings the first head of the Kodokan, the late Kano Jigoro Sensei, asked if I would give a talk regarding Karate. Feeling that I was something of an amateur, I initially demurred, however he was insistent, so I relented saying, "I don't know if I will be able to fully answer all your questions, however if you would like to see a demonstration of Kata…"

I received an answer in the affirmative, so I paid a visit to the Kodokan. In my mind, I would only be demonstrating for the senior members and administrators. As it turned out, I was surprised to find nearly a hundred people were in attendance.

I had travelled to Tokyo alone, so I enlisted the aid of Mr. Gima Shinkin, a native of Okinawa who happened to be studying at the Tokyo University of Commerce.[62] I had trained Karate with him since I was a student and we demonstrated a variety of Kata and Kumite

---

[61] Funakoshi Sensei uses the English word "chance" here.

[62] Gima Shinkin 儀間真謹（1896 ~ 1989）

techniques. Following the demonstration, we had a friendly chat. Kano Sensei asked me, "How long would it take to learn all the techniques?"

"Well, I would have to say it would take more than a year." I replied.

Kano Sensei then asked politely, "That is a bit too long, however I would like to ask if you could teach a few of the fundamental Kata?"

As a person from a provincial area, I was honored and flattered to receive this request from a man of such high standing in the martial arts community.

At last the exhibition ended and, as I was preparing my things to depart, an art teacher by the name of Kosugi Mizui came to visit. These days he is known by the name Hoan. He explained his reason for visiting, saying something along the lines of, "I have travelled to Okinawa in order to find subjects for my art. Since then I have been extremely interested in Karate, however there are no instructors in Tokyo and no books on the subject. I was wondering if it would be possible to receive instruction from you?"

So, several times a week I began going to the house where Hoan Sensei sponsored a drawing and painting group called Tabata Poplar Club and taught the members Karate.

After going back and forth to the Tabata group, the kernel of an idea began to form in me, and it grew in size. I thought, "If I head back to Okinawa now, I have no idea when people on the mainland will be able to practice Karate. Is introducing Karate to the mainland...something I can do?"

So, after discussing the matter with my Senpai back home, I moved into the Meisho Juku, which was a dormitory for students from Okinawa in the Koishikawa section of Suidobata, Tokyo. There, I borrowed a lecture hall to use as a temporary Dojo.

However, though I was the Shihan, head instructor, I didn't have a penny to my name and could hardly ask for money to be sent from my homeland. Ideally, if I could obtain funds from supporters of my project that would solve the problem, but finding the rare person willing to offer support to such an endeavor was clearly beyond the realm of possibility. The main issue was in this era, very few people had ever heard of Karate.

Though I was staying in a dormitory and was classified as a student, the rent, while cheap, had to be paid. Thus I had no choice but to get qualified as the front desk clerk, a job that also included being the watchman and gardener for the dormitory. Thus I settled myself in to a three Tatami mat sized room that served as both the front desk and office to the dormitory.

My work consisted of sweeping out both the inside of the dormitory as well as the outside, doing gardening, receiving and distributing mail for students along with any other odd jobs. In the end, I was quite busy. I conducted Karate lessons in the lecture hall whenever I found time and the students weren't using it.

That being said, I knew exactly how few people were going to attend each training session, meaning my grand desire of spreading Karate felt like one person trying to do Sumo wrestling.[63] Despite this I remained firm in my resolve.

However, the issue of my poverty had not been solved. While my rent was covered by doing odd jobs at the dormitory, I still had to pay for my meals. If I recall correctly, the total cost for meals for the month only came to about six or seven Yen, but as I had very few students joining my Dojo, even that was difficult to scrape together. Actually, it is probably more accurate to say that I was chronically late paying. Finally, I resorted to convincing a cook to take Karate lessons. After strong-arming him into attending class, he then waved my built up meal fees and gave me a future discount. While this was the most difficult period of my time in Tokyo, looking back on it I realize with fondness it was a fun time.

Truthfully, even in the midst of my struggles, I didn't think it was all that terrible.

Even from a young age I was known as a person who was obsessed with cleanliness. I loved to clean and by doing that work I was enabling myself to continue Karatedo training which I felt was my destiny. In short, it was hardly a difficult situation for me.

There were some interesting stories from that time.

Karate was becoming something of a curiosity and magazine or newspaper reporters sometimes stopped by hoping to write a light piece about training at my Dojo. Once a reporter came by right as I was sweeping out the garden and ended up getting embarrassed.

The man probably saw me from behind with my Kimono tucked up into my belt, away from my legs as I swept and mistook me for a servant, calling out, "I am looking for a Funakoshi Sensei."

---

[63] Funakoshi uses the saying *Hitori Zumo* 一人相撲 which refers to tilting at windmills, but was also a type of street performance where one person mimicked a wrestling match

I answered briefly, then went in the back door, raced to my room and quickly washed my hands and put on my Hakama before walking to the front door and greeting my guest, "I am Funakoshi."

The man stared at me in shock, mouth agape. "I apologize for not addressing you properly earlier!"

I had to chuckle at his response, it is unlikely anyone would see a rather small man sweeping in the garden and presume he was a Karate practitioner. Probably better to say the whole incident was my fault.

Karasu Uri no Yurai, Origin of the Crow Gourd

This happened two or three years after I began living in the Meisho Juku Dormitory.

The house of Baron Yasuo Matsudaira (Princess Chichibu's foster parents) was next door to the dormitory. One day, their butler appeared at the front door, "I am here on a small task, regarding the elderly caretaker of this facility. He has have been cleaning the sidewalk in front of our house every morning for some time, and I would like to present to him this small token of our appreciation. Please also tell the elderly caretaker that the lord of the house also extends his thanks and appreciation."

After finishing his polite greeting, the butler handed me a box of sweets. Since the people of the Matsudaira household only saw me sweeping very early in the morning, they had no idea I did anything else other than that.

In the end, I got a fine box of sweets to enjoy. A few years later, the butler came by again, saying, "I have to say that I only just realized, after several years, that you are not an *elderly caretaker*. I do apologize. I ask that you please forgive me." The man said, before lapsing into silence.

Local children also stopped by.

Children would often play in the yard that surrounded the Meisho Juku. Some of the rowdier boys would leave scraps of paper around despite the fact that I had just swept the yard clean. They would also knock leaves off trees, leaving me with more things to clean. It got to the point where I had to give them a warning.

"You young fellows, play without leaving bits of trash around if you please!"

However, one time there was a very rambunctious young boy who had a particularly lively vocabulary and when I chided him he shouted in a rapid staccato, “I’m an old man, I am an old man that looks like a crow gourd! I am a crow gourd, a crow gourd!”

The other kids around immediately jumped on the bandwagon and began the same chant resulting in a riot of noise. But what did they mean by “crow gourd?”

Later, still finding this rather mysterious I looked at myself in the mirror and suddenly burst out laughing. It was just as the children said, I look just like a crow gourd! Even though I don’t drink Sake or smoke, over the years my face had taken on a rather unbecoming redness. As I was younger then, my smooth skin with red coloring meant I looked exactly like the coloring of a crow gourd. After that the children didn’t know me as the “elderly caretaker” or “Karate Sensei” but as the “crow gourd.” I always though this name they called me was rather interesting. Humans, when stripped naked are all the same. Once again I am reminded of the phrase,

空即是色, 色即是空

Kusoku-zeshiki, Shikisoku-zeku

Emptiness is form (color,) Form (color) is emptiness

The same applies to me being referred to as crow gourd, and I for one was delighted with the name.

That being said, I remained ever poor. Thanks to the cook I was able to cover the cost of meals, however that era eventually came to an end and I needed to start paying again.

One day I decided to part the Noren, entrance cloth, of a pawn shop. While this was a tough decision for me, I doubted anything I owned would bring much money. I was nervous going into the pawnshop because I might run into a student from the dormitory or someone from my neighborhood.

In fact all I had was an old derby hat that had small holes in it, hardly something worth more than a trifle. I handed it over to the clerk and waited for him to open his mouth and tell me how much it was worth. The clerk made a dubious face and carried it into a back room of the shop, where I could hear him whispering to another man who I presume was the proprietor of the shop. Finally, the clerk returned and, placing money on the counter said, “Will this be acceptable?”

I don’t recall the exact amount, but it was far above what I had expected. Later I learned later that the clerk’s younger brother studied Karate at my Dojo.

I was almost in tears with thanks for that kindness. Thinking back, I must admit that it was thanks to the kindness of such people that I was able to make it through those early years.

Kosugi Sensei along with the Poplar Club members were kind enough to donate a Yukata and a pair of Geta sandals. Even now I feel brightened recalling their kindness.

Mebae no Koro, Beginning to Sprout and Grow

At first one person would show up at the Dojo, asking to learn about Karate, then two, then at some point I went from five students to ten, then gradually numbers increased to fifteen.

For the most part many of them stopped by the Meisho Juku Dojo on their way home from work, to sweat through an hour or two of training. All of them were serious and deeply enjoyed Karate.

Kasuyo Mahiro (wearing a suit) Funakoshi Gichin (to his right) with the Keio University Karate Club. 1932

They, in turn, inadvertently became the lead advertisers for the Dojo. This is because even people who didn't really understand what Karate was all about would be able to answer, "Karate? Ah, down that road," in response to a query. Soon the number people unfamiliar with this art began to decrease. Further, at some point word spread amongst the university students about Karate. One day Kasuyo Mahiro, a German Professor at Keio University, along with two or three other professors and students visited the Dojo. They said they wanted to study Karate.

Around this time the outlook was gradually lightening from what seemed to be pitch blackness into something much more positive. Day by day the Dojo, which up until now had been a bit lonely due to the lack of students, became brighter and livelier. This served to show that while I was living in Tokyo I was no longer having to "do Sumo by myself."

Shortly thereafter, the students, teachers and other school employees at Keio University founded a Karate training group which signaled the beginning of Karate being taught at universities on mainland Japan.[64]

Following that, students from Takushoku University, which was near the Meisho Juku dormitory, began to appear at the Dojo in regular succession, no doubt due to its proximity.

Thus, in addition to teaching at the Dojo, I was also instructing at the Karate club at Keio University and the Poplar painter's group. In short, I was getting quite busy. One day, a well-dressed gentleman appeared and asked to watch training. I immediately led him to a spot to watch and a few days later he returned, escorting the scion of a noble family, eagerly requesting to join the Dojo.

This young nobleman was formerly a marquis and current member of the House of Representatives, Saigo Kichinosuke. I believe at the time he was enrolled at the Peer's School.

He was an incredibly serious student even going so far as to move into the Togo Residence hall as it was closer to the Dojo. I was surprised at the passion he had for training.

---

[64] Funakoshi Sensei uses the word *Koshi* 嚆矢 referring to a whistling arrow used to signal the start of battle but can also refer to a new dawn, or beginning of something.

Translator's Note: *Koshi* Whistling Arrow

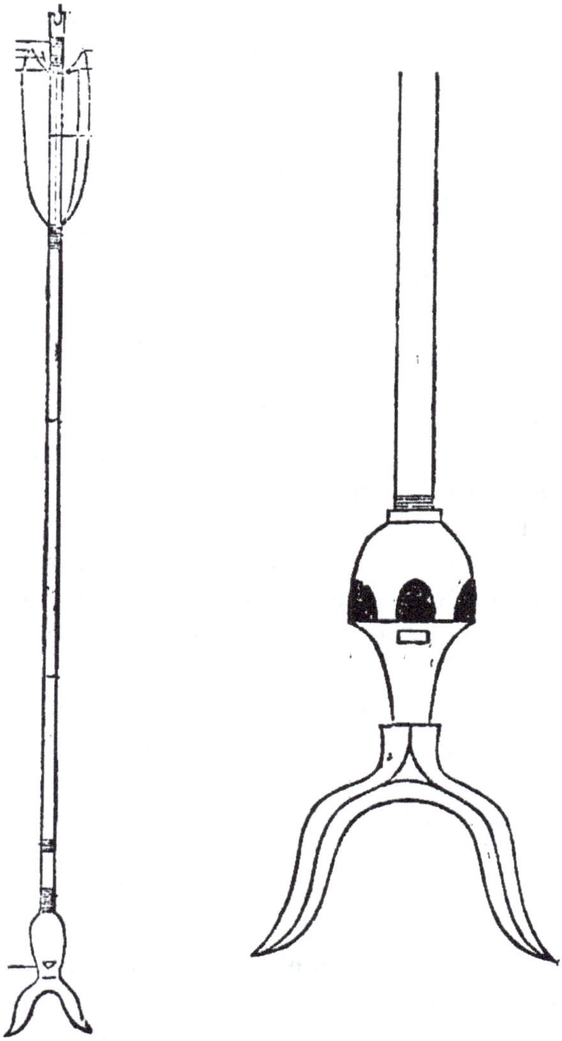

Illustration of a *Koshi* or *Kaburaya*, whistling arrow with a Karimata, goose thigh, arrowhead. *Japanese Armored Horseback Archery Research Society* 日本甲冑騎馬研究会 1924

When shooting on horseback of primary concern is the ease of nocking an arrow. For that reason the nock on the back of the arrow differs from Kyudo, Japanese traditional archery. If you are looking for ease of nocking an arrow with two feathers is best. Three feathers will allow the arrow to rotate. Four feathers will prevent rotation.

Right away I went to the owner of the Togo Residence Hall and told him that a young master from an aristocratic family was staying at his facility, he was completely shocked. Since he felt the main building was not clean enough for someone of that standing, he arranged to have the young lord moved to the Myogadani residence hall. There Saigo lived for several years, commuting to the Peer's School and my Meisho Juku Dojo.

From then on the increasing number and variety of students attending the Dojo could be judged by the differing patches on university uniform hats.

There were students from Waseda, Hosei, Nippon Medical College, the First Higher School, Tokyo Imperial University, the Tokyo University of Commerce and the Tokyo University of Agriculture. Day by day I felt my chest swelling with pride.

At the same time, Universities were starting their own Karate clubs. One was formed at the Nikaido College of Physical Education. Further I was invited to give Karate seminars at both the army and naval academies. I was also receiving more and more requests for lectures. Thus, it was no longer necessary for me to do gardening, or sweep the halls of the dormitory or any other such minor jobs. I was delighted to be so busy.

It goes without say that as the number of students at the Dojo increased, I had to do more than just teach and travel out various university clubs to instruct.

I frequently spoke with the parents or older siblings of students who came by the Dojo to thank me for helping their son or brother become physically and mentally stronger. Sometimes I would receive several of these visits a day.

I believe this came about mainly to due to the self-discipline of the students, but there were, in fact, some cases of university students being against the formation of Karate clubs at their universities. I believe it is due to the Karate students' self-discipline that they were able to dispel their fellow student's misconceptions about Karate and change their perception of the art. This happened more than two or three times.

Thanks to this, I no longer had to sell goods to the pawn shop. However, one day the proprietor of the Edogawa River pawnshop came to my Dojo. "I haven't seen you at my shop in some time and I was afraid some illness had taken you. But I see you are quite healthy, which is the most important thing!"

He never said anything about "hoping to see me at his shop again," but he did leave a large box of delicious sweets, which I was quite embarrassed to receive as I was in no way ill.

Mysteries of Karate Recorded in Print

It was not long after I arrived in Tokyo that Kosugi Sensei suggested, "Wouldn't it be a good idea to write a reference guide to Karatedo?"

So, I began by asking people from my homeland, starting with Asato Sensei and Itosu Sensei, for any information they had as I was hoping to gradually build a collection. However, based on the replies I received none of my Senpai or friends had any written materials on Karate. In short, nothing substantial had ever been written about Karate.

In the end, the base of my book was comprised primarily of what I had heard when I was training in my homeland in my youth as well as the instructions and corrections I was given by my Sensei. Finally, in Taisho 11 (1922) my book *Karate: Ryukyu Kenpo* was published by Bukyosha.

I was honored to have many eminent gentlemen write forwards or contribute calligraphy. Among them were Marquis Sho Sho, of the family of the former lord of Ryukyu Domain, Admiral Yashiro Rokuro, Vice Admiral Ogasawara Chosei, Count Goto Shimpei, Lieutenant General Oka Chiyomatsu, Rear Admiral Kanna Norikazzu, Professor Tonno Norihiro and Sueyoshi Bakumonto of The Okinawa Times among others.

However, not only was this was the first book I ever wrote, but it was also the first book on Karatedo. Now when I re-read it I am ashamed at how incomplete it seems, but at the time I really poured every part of myself into the task of writing it. For that reason it is the book that really stays with me the most.

I divided the book into the following five chapters: *What is Karate? The Value of Karate, How to Train Karate and How to Teach Karate, The Organization of Karate* and *Fundamentals and Kata.* I also added an appendix wherein I discussed the concept of *Karate ni Sente Nashi*, Karate practitioners never strike first, (this refers to the precautions a person studying this art should take) as well as *How Karate is*

Perceived. The cover was kindly designed by Kosugi Hoan Sensei, who was still going by the name Misei at the time.[65]

Here I would like to reproduce the introduction I wrote to my book. I think it would be a good idea to show how I thought about Karate at the time.

> Within the shadow of culture lie its failings, just as rainy days lie in the shadows of clear days. Our history is full of chaos following order, and then reversing, decline following prosperity then reversing, as if it were a rule. The heavens and earth follow the natural and logical cycle of Yin and Yang, light and darkness, sun and moon, male and female. It follows then that you should simultaneously study both the literary arts and the martial arts without considering them to be separate.
>
> If you trust that the fair weather of today will continue tomorrow, you will not be able to escape the terrible effects of floods and storms. Thus, it is essential that you prepare yourself.
>
> Even in times of peace you must not forget the upheaval inherent in war, and to use times of peace to train your body. This is the mentality and the defining characteristic of the Japanese people. Something that is unparalleled elsewhere is how Japanese deliberately do not wait for what others recommend and instead take the initiative. I have come to recognize this as something naturally expected of us Japanese.
>
> These days we are enjoying an era of peace, and our country is making great strides forward in every field. Swords and spears are now being stored away as we head into fall[66], and nowadays we forge our bodies barehanded and without weapons. It has become clear that we are entering an era when people are calling for the culturally refined martial art Karate, "empty hand" sometimes referred to as Karate "Chinese hand" that can be used for self-defense.
>
> Whenever I meet people they invariably find a chance, or a moment to bring up the same thing in conversation regarding Karate "Chinese hand" or Karate "Empty hand." They ask

---

[65] Kosugi went by the name Misei 未醒 until after the Russo-Japanese war, when he worked as a news reporter. Afterwards he became famous for his anti-war literature.

[66] Funakoshi Sensei's book was published in November of 1922

directly, "Are there really no documents? Is there not a book on the subject?" Other times people living in remote places asked if there was some sort of correspondence course available and other such requests. Further, the results of physical fitness tests administered by the military to determine young men's suitability for service have been declining year after year. Thinking the moment was right for something to encourage the preservation of what could only be considered an art created entirely in our nation, I took up my pen. Yet as there was not a single resource that could be used as a reference, I threw my pen aside.

However, the urgent demands of the times have now made these reasons no longer an excuse for me to avoid the effort of writing. This time, I find myself in a position where I must hasten a compilation describing Karate at all costs. While this is the origin of my book I would be hard pressed to say it is the definitive work on this art.

Ryukyu Kenpo: Karate received praise from all parts of Japan and, four years later in Taisho 14 (1925) was expanded and re-issued by Kobundo publishing under the title of *Renten Goshin Karate Jutsu, Strengthening Your Body and Mind Self-Defense Karate.*

Later, after I changed Karate "Chinese hand" to Karate "Empty hand," improved the training methodology and organized the Kata I released *Karatedo Kyohan, An Instructor's Guide to Karate* in Showa 10 (1935.) Kosugi Hoan Sensei also designed the cover of this book. I am currently working on an expanded edition if I can whip these old bones into action, and it should be available soon.

There were other people reporting on Karate from the outside such as the magazine *Kaizo, Restructure.*[67] The reason is because the founder and president, the late Yamamoto Sanehiko, was a native of Kagoshima Prefecture, and thus Okinawa was nearby and that piqued his interest in Karate.[68] Just by coincidence a student of mine, Miyagi So, happened to be working in the editorial department and I have not forgotten all his hard work.

---

[67] The article appeared in the July 1935 issue of *Kaizo* 改造 This was a Japanese general-interest magazine that started publication after the first world war

[68] Miyagi So 宮城聰 (1895~1968)

There were other publications, but I have to regretfully say that most of them did not describe the truth about Karate. For example the following excerpt, which I included in my book *Ryukyu Kenpo: Karate*, in the appendix, *How Karate is Perceived,*

> In the pages of the *Tokyo Nichi-Nichi Newspaper*[69] there was an article with the large caption, *Karate: The Mysterious Martial Art of Ryukyu*. This is an outline of what it contains:
>
> While the main purpose of Karate is to build a strong body it also serves as an art of self-defense. A person who has done extensive training in this art can leap up from a seated position and kick the ceiling, shatter a piece of green bamboo with his fist, split two or three thick boards stacked on top of each other, cut clean through two pieces of *Hosobiki* general purpose rope, carve a stone signpost with his finger, climb around roof beams with just his hands and do triangle jumps. In short many mysterious techniques that seem far beyond the realm of what a normal man can do.

Clearly none of these things are outside the realm of possibility, but to describe them mysterious and beyond the capacity of what a person can do is absurd. Unfortunately, this was the prevalent *Karate-Kan*, the way Karate was viewed by the public at the time.

---

[69] Present day Mainichi "Daily" Newspaper

Showa era photo of Admiral Yashiro Rokuro in front of his home

Stories of Friends, Petals Plucked From the Flower of Memory

One of the first military officers to recognize the value of Karate was Admiral Yashiro Rokuro, the naval officer who viewed a Karate demonstration when brought his fleet into Okinawa and was so impressed that he ordered his officers and men to train in this art.

Interestingly, though I have no idea how, Admiral Yashiro learned that I was in the capital city I nevertheless received an invitation to visit his house in the Hara Town section of Koishikawa, Tokyo. Feeling nostalgic, I immediately agreed and set off for his residence.

I was delighted to find that the Admiral remembered his visit to Okinawa in detail. He told me that he wanted his children and grandchildren to learn Karate, so I went to his house once a week, and had one class for his children and anther for his grandchildren.

As most of my readers are aware, the Admiral is a hero of the Russo-Japanese War and also an exemplar of morality. Whenever I had a lesson scheduled at his house, he would greet me at the gate, wearing a well-ordered Hakama, and following the practice sessions, he would personally walk me to the gate.

Whenever we chatted, I would hear stories of his experiences as a naval commander which invariably contained profound lessons. The distinguished way he relayed such experiences influenced me both as a person and as a Karate practitioner, something for which I am profoundly grateful.

I also learned many valuable things from another naval officer Takeshita Isamu, who later in life became an admiral. In addition, and this may sound strange, but a number of Sumo wrestlers who I happened to be friends with as well.

While my younger readers may not be familiar with his name, Yokozuna Onishiki Uichiro was one of such Sumo wrestlers who visited the Meisho Juku Dojo with some junior members of his Sumo Stable.

However, as the Dojo was a Tatami floored room typically built for use as a lecture hall, it was wildly inappropriate for men of that size to train. In the end, as it was doubtful the floor could withstand the weight of several such Sumo wrestlers, I went to the Sekitori Sumo Dojo in Ryogoku and taught Karate there.

Onishiki would always come and train with the younger wrestlers, or, at times, observe their training with such intensity that I have to admit to feeling the need to offer a bow in respect to the passion he devoted to his juniors martial arts progress. Onishiki retired soon after and opened a restaurant in Tsukiji. He invited me to dine once and when I arrived he offered me a Zabuton cushion to sit on while he himself sat directly on the Tatami mat floor, clearly demonstrating the fact that he was the student and I was the teacher. At the time I was rigidly formal, as the situation dictated so I did not mention it, however I was deeply impressed with his adherence to traditional politeness.

Fukuyanagai, who unfortunately passed away due Fugu poisoning, was another Sumo wrestler who frequented my Dojo. [70] At the time, Sumo wrestlers would frequently travel around the country for tournaments, so they were rarely in Tokyo. I am not sure if it was because they wanted to make up for time they missed, but whenever the Sumo wrestlers had time off from travel matches, or were not participating in a tournament, they would train Karate with ferocious intensity.

Further, whenever they returned from a tour to other parts of Japan, they always came to the Dojo to pay their respects, inform me of their return and relate the events.

One time, I was walking with Yokozuna Onishiki near Ishikiri Bridge in Koishikawa when it began to rain. While he was fortunately carrying a Western-style umbrella, I had forgotten mine.

Onishiki hastened to open his umbrella over both of us, but as he stood over 6 Shaku, 6 feet, tall and I was a "small soldier" standing only 5 Shaku, 5 feet, tall, his generosity clearly wasn't keeping me dry. He suddenly said, "Sensei, take this!" And handed me his umbrella, while the Yokozuna calmly spread a Tenugui cloth over his own head as we walked.

In addition to Onishiki and Fukuyanagi, I also taught Karatedo to three or four other Sumo wrestlers. However, though I was instructing

---

[70] Fukuyanagi Isaburo (1893~1926) died from eating Fugu that was apparently not fresh. After eating the fish he was wracked with intense abdominal pain, and lapsed into a coma. He was treated in the traditional manner of being made to drink soup made from the boiled nest of a barn swallow and bathed in cold water. Though he briefly regained consciousness, he passed away.
The toxicity of Fugu varies among species, from fish to fish, the part of the fish, as well as the season and area caught. The poison does not break down even when heated to 600° Fahrenheit.

them, I felt that I, in turn, learned a lot from them. Whenever I recall teaching those famous Sumo wrestlers who dominated the sport, I reflect that, in the end, Sumo-do and Karatedo follow the same philosophy.

In addition there is the late Count Goto Shinpei, who served as the first president of the support society, former Aide-de-camp to the Emperor of Japan Baron Nara Takeji, Admiral Yashiro Rokuro, General Takeshita Isamu, Mr. Tokonami Takejiro as well as many others.

As for members of the Dojo, there are distinguished gentlemen from every industry and profession represented, including, from the art world, Kosugi Hoan, from the realm of politics there was former Ministry of Labor Mr. Kosaka Zentaro, Mr. Saigo Kichinosuke, from the calligraphy world there is Mr. Uemura Wado as well as Togawa Yukio from the literary world as well as many others. There are a great many others that I wish to write about but due to space considerations I will have to wait until another time.[71]

Nonsense Regarding Why I Wear Hoba Geta[72]

Though I ostensibly travelled to the capital just for the physical fitness exhibition, my stay had been extended considerably. However, since I felt that my work of introducing Karate to the mainland had only just begun, I decided I couldn't leave Tokyo until this goal was achieved. Nay, I felt I couldn't return to Okinawa until I had laid a firm foundation, as if my own bones were embedded in the soil of the mainland. I thought, "That seems like exactly what my goal should be."

---

[71] Goto Shinpei 後藤新平 (1857~1929)
Nara Takeji 奈良武次 (1868~1962) Army General
Takeshita Isamu 竹下勇 (1870 ~ 1949) Navy Admiral and also a patron and practitioner of Judo, Sumo and Aikido.
Takejiro Tokonami 床次竹二郎 (1866 ~ 1935)
Kosaka Zentaro 小坂善太郎 (1912 ~2000)
Uemura Wado 植村和堂 (1906 ~ 2002)
Togawa Yukio 戸川幸夫 (1912 ~ 2004)

[72] Funakoshi Sensei uses the English word "nonsense."

So, at some point, I began to get settled into my new life on the mainland. Therefore, I began to think about bringing my family up to Tokyo. Fortunately, my eldest son had moved to the capital before I did, and my second and third sons all eventually joined me. Once my family was all together, I began to feel more confident about my goals. I wrote a letter to my wife back home. However, Okinawa is a land where everyone takes seriously their responsibility to worship their ancestors, add to this the fact that my wife was a very devout follower of Buddhism meant that she was adamantly against moving the graves and Buddhist mortuary tablets to another location. Thus, the situation and the circumstances did not enable me to move my entire family up to Tokyo as I had originally envisioned.[73]

In short, the answer I received from my wife was that the graves of my ancestors were not only guarding me but also the source of my enlightenment and zeal for work.

As it turned out my devotion to work was tested after the Great Kanto Earthquake that struck on the first day of September in Taisho 12 (1923.) Fortunately, the Meisho Juku Dojo escaped damage. I volunteered in town cooking rice for those in the community who had to flee the calamity in addition to working with the local community protection unit. However, many members of my Dojo who met with misfortune and were either killed in the disaster or were missing. It still pains me to think about all those people. To make a long story short, the earthquake and ensuing fires meant that everything in Tokyo ground to a stop, so obviously there was no Karate training going on. For the most part, all the members of the Dojo were hard at work just to enable everyone to eat. I was, of course, doing the same thing.

The first thing we had to do was clear the rubble from the buildings that had burned down along with transporting the bodies of the dead. At some point I was introduced to an offical from the Number One Sogo Bank in Kyohashi and I, and all the members of my Dojo, were hired as a group for day-labor to copy documents. I believe there were around twenty students from the dormitory. They, along with about ten members of the Dojo, which made about thirty of us, walked every day from Suidobata to Kyohashi.[74]

---

[73] One other possible reason for Mrs. Funakoshi's hesitation may have been tattoos. The tradition of all women tattooing their hands did not end until the beginning of the twentieth century.

[74] The distance was probably around four miles each way.

I can't recall how much we got paid each day, but I do clearly recall that the hours were extremely long.

At the time no one was wearing shoes. Everyone had on Geta, wooden sandals, or Waraji, straw sandals. As for myself, I wore Hoba Geta.[75] No doubt many thought that as a "small soldier," or man of short stature, I was wearing them out of a sense of inferiority, however, ever since I was young I have worn Hoba Geta in order to train my legs.

Sometimes you see Karate practitioners wearing iron Geta as a training device, however I wear Hoba Geta as my daily footwear in order to strengthen my legs and feet. That is why I selected Geta made of various types of wood that, when combined, are very heavy. That is why I never change my footwear no matter what banquet or event I am attending.

Though I am well into my middle years I still walk wearing my Hobaku Geta, which make a clattering sound as I walk, just like the youngsters. Thinking back on that time it seems like it was a nonsense decision; no at the time it must surely have seemed bizarre, but as I considered it part of my training, I wore Hoba Geta everywhere I went.

---

[75] Hoba or "Japanese hackberry-toothed sandals." The wooden stilts on the bottom of Geta are referred to as "teeth." These types of sandals were popular with students and students who lived in people's houses and worked part-time as a butler to pay for rent.

Funakoshi Sensei practicing Makiwara in Hoba Geta

Birth of the Shotokan

The Meisho Juku got its name from the fact that it was build right at the end of the *Meiji* Era and beginning of the Tai*sho* Era, however since then no repairs had been done. Finally, the prefecture as well as the Okinawa Society for Educational Development arranged a renovation.

So, though it had been my home for a long time, I would have to leave for a time. Fortunately, Kendo practitioner Mr. Nakayama Hiromichi, with whom I had been on friendly relations with for several years, heard of my situation quite by chance and graciously offered the use of his own Dojo, located in Hogo Yumicho, when his Kendo training was not taking place.

Nakayama's Dojo was said to be the greatest in Japan, it was a magnificent building constructed with a gable roof like a castle, and could in no way be compared to the Meisho Juku Dormitory Dojo. Due to this, the number students that came to my training increased steadily. In addition, students of Nakayama Sensei also started training with me, meaning that I finally began to feel that my work at expanding Karate was beginning to bear fruit.

As for myself, I was able to find a house to rent near the Dojo that was about 3 Ken, 320 square feet. However, soon after I was able to locate another rental property that had a larger garden for doing training in the Masago Town area. With that, it seemed I was finally moving away from a life of poverty.

As it turned out, though Nakayama Hiromichi had generously allowed us to use his Yumi Town Dojo, the number of students at his Dojo was increasing and, somewhat belatedly, the number of students studying Karate was also increasing.

With that, the previous arrangement of running Karate lessons when Kendo was not in session became untenable. I was now worried that holding Karate training at Nakayama's Dojo was becoming troublesome for the Kendo Sensei and felt that I should take action. Though I pondered two or three concepts, it was beyond my ability to plan and construct a Dojo. As it turned out, in the eleventh year of Showa (1936) I believe it was, Dojo members formed a Dojo Construction Committee and, at long last, we were on track to have a primary training facility.

Left: Nakayama Hiromichi 中山博道 (1873~1958)
Right: Nakayama Zendo 中山善道 (1900~1981)

Nihon Kendo to Seiyo Kengi

日本剣道と西洋剣技

Japanese Kendo and Western Sword Techniques

The committee presented their plan to devotees of Karate, who at this point lived all over Japan. The Dojo students, despite extraordinary hardship, began to steadily collect donations until sufficient funds were available to construct a Dojo in the Zoshigaya, Toshima section of Tokyo. This happened in the early spring of Showa 13 (1938.) At that time, there was no official Karate Dojo anywhere, thus it is not an exaggeration to say it was a new dawn in the era of Karate.[76]

Hanging at the entrance to the Dojo a large sign that read Shotokan, a name which the Dojo as a whole agreed upon. The name Shoto is the pen name I used when I was studying Chinese literature as a young man. I would never have imagined that one day that name would be used for a Dojo. Even now it is impossible to put in words the emotions I felt when that signboard was hung over the Dojo. I thought, "Asato Sensei, Itosu Sensei, today a marvelous Dojo has been built!"

That day, I lit incense in my room and joined my hands in thanks as I prayed to the spirits of my beloved masters, informing them of these events. At that moment I felt as if my two great teachers were smiling at me and saying, "Good work, Young Funakoshi, good work!"

However, they also cautioned me, "But, Young Funakoshi, it is too early to rejoice, you still have a lot of work to do. Today, Young Funakoshi, is when things actually start![77] It is too early to celebrate!"

However my true wish was for my two beloved teachers to have had a chance to see the new Dojo. It would have been a great honor to have the two seat themselves in the *Shihan no Seki*, spot reserved for head instructors. Alas, the two Sensei, who when seated project nothing but compassion, are no longer in this realm. They have set off on their journey away from the mortal coil.

Now that the Dojo was open, I had a chance to work out the teaching schedule as well as a curriculum. I also established rules for both Dan and Kyu ranks around this time.

---

[76] Funakoshi Sensei again uses the word *Koshi* 嚆矢 referring to a whistling arrow used to signal the start of battle but can also refer to a new dawn, or beginning of something.

[77] The spirits of Asato Sensei and Itosu Sensei use the English word "start" here.

New students joined the Dojo day by day. However, by that time I was past sixty years of age. I was determined to continue on living. My feeling was that my real work started from this point.

Fortunately, I looked young for my age and was able to move in a way that belied my years. Nowadays, that is probably the only strength I have. I threw myself into training with renewed passion, surpassing even what I had before.

It may have been due to my rejuvenated teaching, however the new Dojo, which at first felt so large, had, at some point begun to feel too small. Further, the vast majority of universities in Tokyo now had Karate clubs, so I had to go and teach at those Dojo as well as my own. In the end, even dividing the sessions into morning and evening became more responsibility than one person could manage.

As a result, from among the graduates who had been members of the Karate club, I selected those that were able to take sufficient time off and designated them *Shihan-dai*, Assistant Instructors, and they took over the day to day teaching of Kohai, their juniors.

As far as the Shotokan Dojo goes, I enlisted the aid of my third son Gigo to take over primary teaching duties, while I worked for the most part to supervise the organization as a whole.

Eventually, not only members of the Shotokan Dojo, but also former members of university Karate clubs began to find jobs in different regions of Japan, moving farther and farther from Tokyo. They, in turn, founded Dojo in those areas as well and sometimes companies would start clubs for their employees to train.

Thus my pilgrimages to regional areas increased in frequency until I began to delegate those duties to senior members of the Shotokan Dojo.

My Enjoyment of Solitude and My Pen Name

People often ask me about the meaning of my pen-name Shoto, Pine waves, I usually say, "Really, it's nothing more than a pen name, there is no special meaning."

This may seem like I am avoiding the issue, but that is, in fact, the truth.

In short, it is a pen name that I have been using for decades, to the point that when I write it on something I almost feel it is my true name.

Now, when asked about it, I really have to think back to how I came up with it. Thus, this chapter will touch on the origin of my pen name, while relaying a memory of my younger days.

The town of Shuri is a *Joka-machi*, a town that developed around the castle of the Ryukyu king. I don't know if it has anything to do with the development of the area, but mountains that surround the castle are covered with Ryukyu pine trees and other subtropical trees and foliage.

Among them is the peak of *Torao-San*, Tiger Tail Mountain, which is surrounded by a particularly dense pine forest. This mountain belongs to the family headed by Baron Ie Chojo, who was one of my first patrons after I arrived in Tokyo. The reason it is called "tiger tail" is because the ridge of the mountain is long, thin and covered with pine trees, resembling a tiger's tail. When viewed from afar it looks like you could reach out a stroke the tiger's tail.

I am not sure exactly when, but I got in the habit of strolling around Mount Torao. Whenever I had free time I would simply wander around this area. Sometimes it would be at night under a half-moon, other times when only the stars were out, covering all the heavens. A gentle wind would blow, causing the sound of the pine trees rustling to carry throughout the wood. Combined, it can only be described as an effect that gives one a glimpse of the divine, while the sounds of the forest seem to be some mysterious music.

From days long past, poems have been written and songs have been sung in many different styles about the subtleness and mysterious nature of the natural world. I too am likely one of those writers, though I am an average poet who writes poorly composed poems on vulgar and clichéd topics.[78]

Another thing to remember is that I was an only child, who was of frail constitution, and that may have contributed to me being introverted. So, while it might be extreme to say I was *Kodoku-Ha*, one who prefers solitude, my enjoyment at being alone in nature continued and perhaps became part of my personality.

So, after training Karate with such intensity that sparks seemed to fly, I found I wanted to indulge in the opposite: stroll around in quiet introspection. Be that as it may, I didn't find any contradiction in preferring to go from one extreme to the other.

---

[78] Funakoshi Sensei refers to himself as a *Sanbun-Shijin*, a poet of meagre ability who's rough, amateurish works do not sell.

Edo Era illustration of Ounoyama park from *Eight Views of Ryukyu*
The area has been filled in and is no longer an island.

I have a similar memory from when I was still in my twenties and working as a teacher in Naha. There was an island right smack dab in the middle of the bay that somewhat resembled the shape of the Okinawan archipelago. There was a dense pine forest there are as well and the island sported Ounoyama Park, the premier nature park in the prefecture.

At the time there weren't any houses on the island, just a Zen temple, a shrine to Benzaiten, the Buddhist goddess of speech, the arts, and learning and a lotus pond. As that was all there was to the island, I would frequently go there with my walking stick and gaze at the thick trunks of the pine trees and generally enjoy the natural beauty.

At that point in my Karate, I had just finished the *Shoki,* early era of training, and entered the *Chuki*, middle era, and I had taken to pondering the true nature of the spirit of Karate while on my walks. So the silence, other than the enchanting, almost supernatural sound of wind rustling pines, gave me time to think, "I want to somehow merge elements of the truth held within this elegant silence not only to *Ken no Michi*, the Way of the Fist, but life in general."

Clearly this captures some of the feeling of the young literati that I was.

As I had been like this since childhood, the word *Shorai*, referring to the sound wind makes as it passes through pine trees, had for some reason stuck in my head. Thus when writing songs or composing Chinese poems, I suddenly realized the name Shoto would be an ideal pen-name.

Interestingly, and this is quite curious, as I had used this pen-name for so long, there were some people who didn't recognize my actual name. Sometimes I would have to write Shoto above my first and last name on a document before realization would spark on people's faces. Names are certainly a curious thing.

Two Entries Regarding the War

This happened during the time when dark clouds were gathering over Manchuria, though the main islands of Japan remained peaceful. I participated in a regularly scheduled *Enbu*, martial arts demonstration, before the Emperor of Japan at the Saineikan Dojo,[79] which is located on the grounds of the imperial palace.

I, along with five or six students, demonstrated several Kata before his majesty. The honor and deep emotion I felt at being able to demonstrate before his majesty has stayed with me all my life. When I was young, and walking almost two and a half miles to train Karate in secret, no one would never have imagined a time when I would be demonstrating this art before his majesty the Emperor of Japan.

---

[79] This building is currently the Imperial Guard Headquarters. Martial arts demonstrations and competitions were frequently held at this Dojo, particularly for Kendo and Judo.

On February 20th of Showa 3 (1928,) Head Instructor of Karate Jutsu, Funakoshi Gichin was invited to the Imperial Guard Headquarters Sainenkan Dojo, to conduct a demonstration of techniques and lecture to the police. Various top officials from the Department of the Imperial Household were in attendance and watched the demonstration intently.

-*History of the Imperial Police* 皇宮警察史
By the Imperial Police Headquarters 皇宮警察本部 1976

Saineikan Dojo, where Emperor Hirohito watched the Tenranshiai, Duels observed by the emperor. This event seems to have been held annually at least in the late 1920s through the 1930s.

Being selected to demonstrate martial arts before Emperor Hirohito was a supreme honor, that almost overwhelmed me with emotion. I had at last been given the chance to present Karate technique I had refined and developed over the past half-century, and my audience was his majesty the emperor. Of course, the emperor had seen my demonstration in Okinawa, when he was still the crown prince. However, at that time Karate itself was neither developed nor was it accepted as a Japanese martial art. To put it in stark terms, Karate was thought of as a *Chiho Bujutsu*, a martial art only done in a remote area by a few practitioners. However on this day, Karate was perceived differently. It was a spectacular Japanese Budo.

The fact that it was demonstrated in front of the emperor is a testament to that…and looking back, it reflects the road I travelled. Thus, the demonstration that day affected me deeply, reverberating deep within my chest.

A few days later, I was invited by Grand Chamberlain to the Emperor Chinda Sutemi to an audience and told me that his majesty vividly recalled the Karate demonstration at Shuri Castle and asked if the instructor at the demonstration at the imperial palace was the same person.[80] It goes without saying that I was so stunned by this revelation that I was choked with emotion.

The situation in Manchuria was finally beginning to boil over and a sense that Japan was in a state of war began to permeate the nation. From that point on, Dojo membership grew in leaps and bounds.

With the beginning of the Second Sino-Japanese War (1937) and the Pacific War (1941) the Dojo was overflowing with young men, to the extent that not only was the Dojo full, but also the adjoining garden, meaning some students were practicing out in the street. The sound of boisterous Kiai shouts and the sight of innumerable young men striking Makiwara became a daily event. One of the things I

---

[80] Count Chinda Sutemi 珍田捨巳 (1857 ~ 1929) was a diplomat who served as Grand Chamberlain to the Emperor from 1927 until 1929.

Count Sutemi Chinda, Japan's Grand Chamberlain, is dead after a single day's illness. His death removes from the immediate entourage of the Emperor one of his most sagacious and most trusted advisers, who held a position much more important than appeared on the surface.

THE NEW YORK TIMES
Jan. 17, 1929

worried about was whether all the racket was disturbing the neighbors. On the other hand, the number of students who either enlisted in the armed forces or were drafted increased day by day.

Young men, faces full of intense determination, would get on their knees and bow placing both hands on the ground while saying, "Sensei, I am heading off!" The same *Waka Musha*, young warriors, had, up until now, been wearing *Keikogi*, Karate uniforms, while repeatedly practicing their kicks and punches against an enemy they could not see.

I received letters almost daily from Dojo members on the front. Some of them described, gut wrenching battles that devolved into *Hakuheisen,* hand-to-hand combat, where the men, who were too exhausted to hold a rifle or carry a sword and were ordered to,

Karate wo Motte, Shitosu Seyo!
Pierce through them with your Karate!

Sometimes the students had declared themselves to be "troops of the Karate Unit" and sent me letters filled with questions about training. The number of Dojo members who died during the war are too many to count.

Thinking back on that war where many fine young men who excelled at Karate were lost for nothing makes my heart ache. Whenever I would receive word that one of them had died in combat I would stand alone in the Dojo and recall their vitality as they trained Karate before joining my hands in prayer.

Translator's Note: Enbu at the Saineikan Dojo

Impressions of the Martial Arts Demonstration at the Saineikan Dojo
By Yokoyama Kendo 横山健堂 (1872~1943)
Police Society Magazine 警察協會雜誌
June 1932 #382 6月號

The most interesting demonstration was Ryukyu Karate Kenpo. This author, while on vacation in the south, saw this sort of training. The Ryukyu warriors, with nary a weapon, developed and perfected a weaponless empty hand style of fighting over the course of a thousand and several hundred years. Something that can be used when facing off with an opponent armed with a sword or halberd.

Napolean the first, during his exile on St. Helen, when he heard from sailors about the nation of Ryukyu, a kingdom without weapons or fortified castles, he declared, "I could never imagine such a place exists!" It is just as Napolean said.

Hundreds of years ago the now ruined ancient castles once stood guard, however at present, the natives do not wear swords, and instead all you can find stuffed in their belts are Kiseru pipes!

I saw some old plays from Ryukyu and in them Karate was used to win duels by knocking the Katana out of the hands of the Samurai from Satsuma. Karate is a martial art unique to the people of Ryukyu and seeing its value one understands why it is the pride of Ryukyu.

Nowadays, in the era following the abolishment of swords in Japan, it is clear this Karate, in particular, is something people should study. To think that the tiny islands of Ryukyu have perfected a martial art, is something that should astonish the martial arts world. While Judo has it curved lines, one particularity of Karate is that it is linear. To sum up Karate, it is the sprit to go into battle against a sword with just your bare arms. The sailor who told Napolean about Ryukyu had never seen Karate. If Napolean had asked about that, the conversation may have been much longer.

While there are already books about Karate available, the art of Karate has been steadily rising to prominence on the martial arts stage, and this demonstration served as a signal arrow announcing further advancement. And this doesn't apply just to the flourishing of this art but is sure to have no small effect on the martial arts world as a whole.

The Death of My Wife

Every day, reports of the progress of the war became grimmer and Japan seemed to be moving closer to defeat.

Then, in the spring of Showa 20 (1945) my third son, Gigo, had to be hospitalized.[81] I had already evacuated to my eldest son's house in Koishikawa-hayashi Town in northern Tokyo, but shortly thereafter the Dojo was destroyed by an incendiary bomb during a concentrated attack, and all traces of it were blown away.

It all happened in a moment.

Thinking back on it, the Dojo was built due to the hard work and diligence of the members. The connections forged through Karate training developed into a warm friendship that crystalized into action and it was this that led to the success of this project. And what they created was the most beautiful *Mono*, thing.[82] Something that I had worked my whole life for. And it was all lost in one night.

I wept as defeat in war approached.

As the confusion and havoc continued, I was no longer able to remain in Tokyo. Thinking of the food situation, I decided to travel to Oita in Kyushu, to where my wife had evacuated.

However, it turned out that refugee life in Kyushu was more difficult than expected. When I say that the people of Okinawa had evacuated to Kyushu, they had done so in large groups, not with friends and family members. My elderly wife, despite these conditions, had not lost the characteristic vigor that she had from long ago.

My wife and I grew potatoes and other vegetables. We also walked along the beach gathering various kinds of seaweed. However, this life didn't last long. One day, my wife suddenly collapsed.

Having exhausted her supply of vitality my wife never recovered from her collapse and instead passed into the next realm. It was the late fall of Showa 22 (1947.)

Her death was even more tragic when you consider we were due to leave for Tokyo the next day to move in with our eldest son. We had even finished all the packing.

That evening, while she was having difficulty breathing due to her asthma, she asked me which direction faced Tokyo. I told her and she

---

[81] Photographs on the previous page are of Funakoshi Gigo.

[82] Funakoshi Sensei makes a point of emphasizing the word mono 「もの」

turned her head in that direction, bringing her shaking hands together in prayer. Next, she asked me which direction faced Okinawa and when I told her she turned her head in that direction and again prayed.

When she turned to face Tokyo, she was doing a *Kyujo* bow of respect towards the imperial palace, and brought her hands together in order to say goodbye to her children and grandchildren. When she faced Okinawa and brought her hands together, she was offering a last prayer to the graves of our ancestors. These were the final moments of my wife, who supported me in poverty through the long years of Karate training to which I had devoted my life.

I had moved to Tokyo around the time I turned fifty and had been living apart from my wife. Life in my homeland had continued to be one of poverty. Thus, we didn't have as many of the happy times that other married couples in this age had. The entirety of her life was spent taking care of her husband who had embarked on the path of Karate training and raising our children.

In Oita, each village had its own crematorium. I am not sure exactly why, but no one other than people from that particular village could use that facilty. As only villagers could use the local crematorium, *Takoku mono*, those from another domain, had to cross the sea by boat a town called Usuki and use that town's crematorium.

However, as it turned out the village, led by the local youth group, granted permission for my wife to be cremated at the local crematorium, the first time any *Takoku mono* had been granted this exception.

Seeing this, I realized that my wife had gone to great lengths to facilitate our new life in Kyushu. After the cremation I set off with her remaining bones and ashes to Tokyo.

As it turned out, and I have no idea how they heard, but at numerous stations along the way Dojo students met the train and consoled my grieving heart. I was not the only one unable to restrain the tears that flowed.

I now felt that wife's death was not in vain. My wife's death had given birth to all those students who came out to pay their respects. It was this outpouring of sympathy that somewhat alleviated my grief.

Karate Zuidan, Miscellaneous Stories About Karate

This first story is sort of in a similar vein to,

Hanashi Jozu, Tatsujin Nashi

Among those who are good talkers, none are expert martial artists

In recent years, even average people seem to have become alarmed by phrases like *Karate sannen-goroshi* or *Karate gonen-goroshi*, meaning anyone who struck by a karate blow will inevitably die from the effects either within three years or within five years. With more and more people using these phrases, I would like to talk about them.

The phrases themselves are rather savage to be sure, and while it is unfortunate that that savagery is reflected on Karate as a whole, it is not altogether inaccurate.

However, even if you refer to *Death in Three Years* or *Death in Five Years*, this isn't actually saying that you will die three years after being punched, or if you get kicked in a certain way you will die in five years.

What these sayings are actually describing is, for example a fight where you are not able to topple your opponent then and there, however several years later, the effects of the fight lead to his death. In other words, this refers to the effects of your attacks shortening the lifespan of your opponent.

I will explain the underlying principle. For example, if you look at photographs of Karate practitioners doing *Tameshi Tsuki*, test strikes, on boards or roof tiles, you will note that the uppermost ones neither break nor bend, yet all the ones below shatter spectacularly.

Thus if, at the moment of impact, you adjust the power in your fist, while the boards below will crack and shatter, the surface of the board on top will remain unaffected. In other words, the same thing happens when you strike a person. If you punched in accordance with Karate training, even if there is no mark on your opponent's skin, you could have quite easily caused internal bruising.

For example, a person may have ended up receiving such an internal bruise, but think nothing of it at the time. I know many people who have seen or heard about such incidents where, a few years after the fight, the man who was struck begins to feel pain and this leads to a big health issue.

This is what is meant in Karate when saying *Death in Three Years* or *Death in Five Years,* however, much like *Ita-wari* and *Kawara-wari*, breaking boards or roof-tiles, it is an unimportant detail.

Take for example *Ita-wari*. If an average youth practiced with Makiwara, after about a year, he would probably be able to break three or four half-inch thick boards. That being said, even if he succeeded he could not be described as a Karate practitioner or one who knows Karate. In addition, if he were to try and use his skills in a real fight, as the saying goes, *A little martial arts training is a dangerous thing*. All that person has really done is strengthen the skin on the back of his knuckles.

Clearly, nowadays it doesn't seem like anyone is foolish enough to try that, in addition no one misunderstands the purpose of board and tile breaking. That being said, when I first arrived in the capital, in the Taisho era (1912~1926) the police department and other officials concerned with public safety were very worried that someone would attempt to misuse Karate for something bad. This is one such story.

One day, a police chief stopped by with a concern, "If someone is carrying a knife or a gun, we can see right away he has possesses a firearm or a bladed weapon and can arrest him for that. However, a Karate practitioner only has his fists. Sensei, I realize that there are all sorts of people training at your Dojo, however recently there are a great many gangs, scheming to cause mischief, so I urge you to caution your students as you instruct them."

I have to admit that this is good advice. If a designated gang were to use Karate to commit a crime, and it was discovered that I was the person that taught them, you could say that my methods were used to commit crime. A *Seiken,* true fist, would have been twisted into a *Jaken*, cursed strike.[83] My heart froze at the realization that if such a thing were to occur, I would be completely dishonored and would have to apologize to every person in the nation. Fortunately, despite have taught ten-thousand, nay, tens of thousands of students, none of them has fallen so low as to employ Jaken against a victim.

---

[83] *Seiken* 正拳 true fist or true strike
Jaken 邪拳 cursed or twisted strike

Karate ni Sente Nashi

For a long time I have been promoting *Karate ni Sente Nashi*, a Karate practitioner doesn't make the first strike. Early on, whenever I wrote about Karate I invariably included that line, however I have a complaint about it. As I wrote in *Karate: Ryukyu Kenpo* (1922,)

> In our world, you should consider that every insect, even if only an inch long, contains a soul and a will to live that fills half its body. As you move through the world, if you are not humble in your manner of speaking, you will find enemies in any direction you face.[84]
> From days long past it is said that it is the tallest of trees that the wind strikes.[85] However the willow tree bends in face of the wind, which means the power of the wind flows off it. Learning humility and respect are the final stage for Karate practitioners to understanding the beauty of human morality.

One hand, no matter what the situation is, adamantly refusing to strike the first blow is not what I am describing. On the other hand you also should not always be in a state where your mind is focused exclusively on striking first, or making the first move. I always tell people looking to join the Dojo this condition: *No matter what the situation, you are never allowed to use Karate to settle a personal fight or resort to using your fists in an emotional situation.* If they understand that condition, I allow them to join.

---

84 *Issun no Mushi Nimo Gobun no Tamashi* 一寸の虫にも五分の魂
Even the smallest beings have their own wills, so do not make light of them

85 *Takai Ki ni wa Kaze ga Ataru* 高い木には風が当る
Those at the top can draw the jealousy of others.

Frequently, some of the younger men are inclined to craft justifications such as,

Gi wo Mitesezaru wa Yu Naki Nari
義をみてせざるは勇なきなり
Seeing what is right and failing to act means you have no honor[86]

Along with,

Taigi
大義
A Great Moral Cause

However, that "moral duty" [87] is based on what you yourself have decided, based on your own viewpoint. If you were to take a step back from that and look at your position objectively, I would imagine that many people would then realize that perhaps the position they are promoting is not so much of a moral duty.

If you are employing your fists in a situation like *Seeing what is right and failing to act means you have no honor* along with *A Great Moral Cause* then that means it has instantly devolved into a life-or-death situation. If, over the course of your life, such confrontations were to happen again and again, it would make life unbearable. I believe you should instead consider whether or not the situation you are in is indeed a moral imperative requiring extreme action. I believe that if you take the time to consider the situation, then things will go more smoothly.

One thing that greatly comes up with people who train is wondering how skilled you have actually become. Some think they have become sufficiently adept that they seek to test their technique, however I caution that this is the height of folly.

---

[86] This is a quote Chapter 24 *On Governing* from the *Analects* by Confucious.
The Master said:For a man to sacrifice to a spirit which does not belong to him is flattery. To see what is right and not to do it is want of courage.

-Translated by James Legge (1893)

[87] *Meibun* 名分 moral duty or a justification

Once, when I was still using the Meisho Juku Dojo, a Karate student who had only just begun training tried to kick the dog owned by the Matsudaira family next door. He was wearing Geta wooden sandals, but when he kicked, the dog ended up biting him, showing that it was the dog that understood Karate never attacks first. Not really a story to laugh at.

Karate Chop[88]

Though professional wrestling hasn't been around for long in Japan, it is clear to everyone that Rikidozan's string of victories has raised Japan's standing, so we are now on equal footing with the rest of the wrestling world.[89] Further, I believe it is impossible not to respect the extraordinary lengths he went to over the course of his training.

It is said that one of his strongest weapons was the Karate Chop. I wondered if Rikidozan had studied Karate early in his career, and was more than a little suspicious of this. However, he told me that when he was ranked Juryo or Makushita in Sumo, my student Togawa Yukio visited him.

Thus, his interest in Karate predates his switch to pro-wrestling to the time when he was still a Sumo wrestler. I believe this episode demonstrates his passion for furthering his martial arts skills. [90] In short, continuous effort is not limited to pro-wrestlers or Sumo wrestlers but something essential for anyone practicing a martial art.

The only regrettable thing is that I have no idea how the rules in pro-wrestling work. Frankly, I don't like being in crowded places, so I never had a change to watch one of his bouts closely. Rather, I have only seen him two or three times on television, so I am hardly qualified to talk about pro-wrestling, so this passage is based mostly on what I read in newspapers or magazines that discussed his Karate Chop.

The word "Karate Chop" has a bit of a easygoing charm to it, but in the realm of Karate is it known as *Shuto Uchi*, Hand Sword Strike.

---

[88] Funakoshi Sensei uses the English word "chop."

[89] Mitsuhiro Momota (born Kim Sin-rak) (1924~ 1963) was a Korean-Japanese sumo wrestler and later professional wrestler. He was known as the father of professional wrestling in Japan. Born in what is now North Korea under Japanese occupation, he moved to Nagasaki to become a sumo wrestler. He competed in Sumo until 1951.

[90] Funakoshi Sensei uses the English word "episode."

手練の手刀が斬り込む「空手打ち」の連打

Cutting in with a well-honed Shuto, repeated Karate Uchi strikes

Rikidozan Album 力道山アルバム
Mitsuhashi Kazuo 三橋一夫 1955

This is because you make the strike by bringing all five fingers of your hand together so your hand looks like a sword. However if you look at the Kanji that make up the word Shuto, Hand + Katana, it brings to mind something along the lines of a toy *Shinai,* bamboo training sword, that children use to play Chanbara.[91] However, there is nothing playful about the Shuto in Karatedo. Just like with punches, the Shuto strikes like a real sword.

[91] Chanbara is play sword fighting.

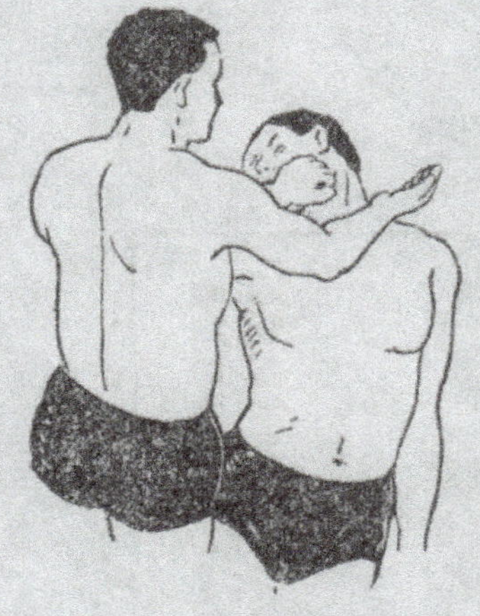

空手打ち　手の側面を縦に使って（これを手刀という）相手を烈しく打つ。

Karate Uchi
Use the side of your hand to strike your opponent vertically from above with a ferocious blow (this is called a Shuto)

Rikidozan and Endo Kokichi: Kings of Pro Wrestling
力道山・遠藤幸吉：プロ・レス王者
Gunji Nobuo 郡司信夫 1954

Striking with Shuto to the veins in the neck, or where the neck meets the torso, will cut a person's breath off at the root and if the strike continues on to the collar bone, that will break as well. Further, all varieties of Shuto contain the same power as the blade of a Katana. You can extend your fingers to pierce, strike with the side, or strike with the back of your hand in an Ura Uchi, backhand. No matter which way you strike, it will still be able to inflict the same damage.

You can also use Shuto strikes for *Tameshi Wari*, test breaking, either boards or roof tiles, however technically speaking it is more difficult to do than with the fist. Clearly a Karate Shuto is vastly different from a Karate Chop.

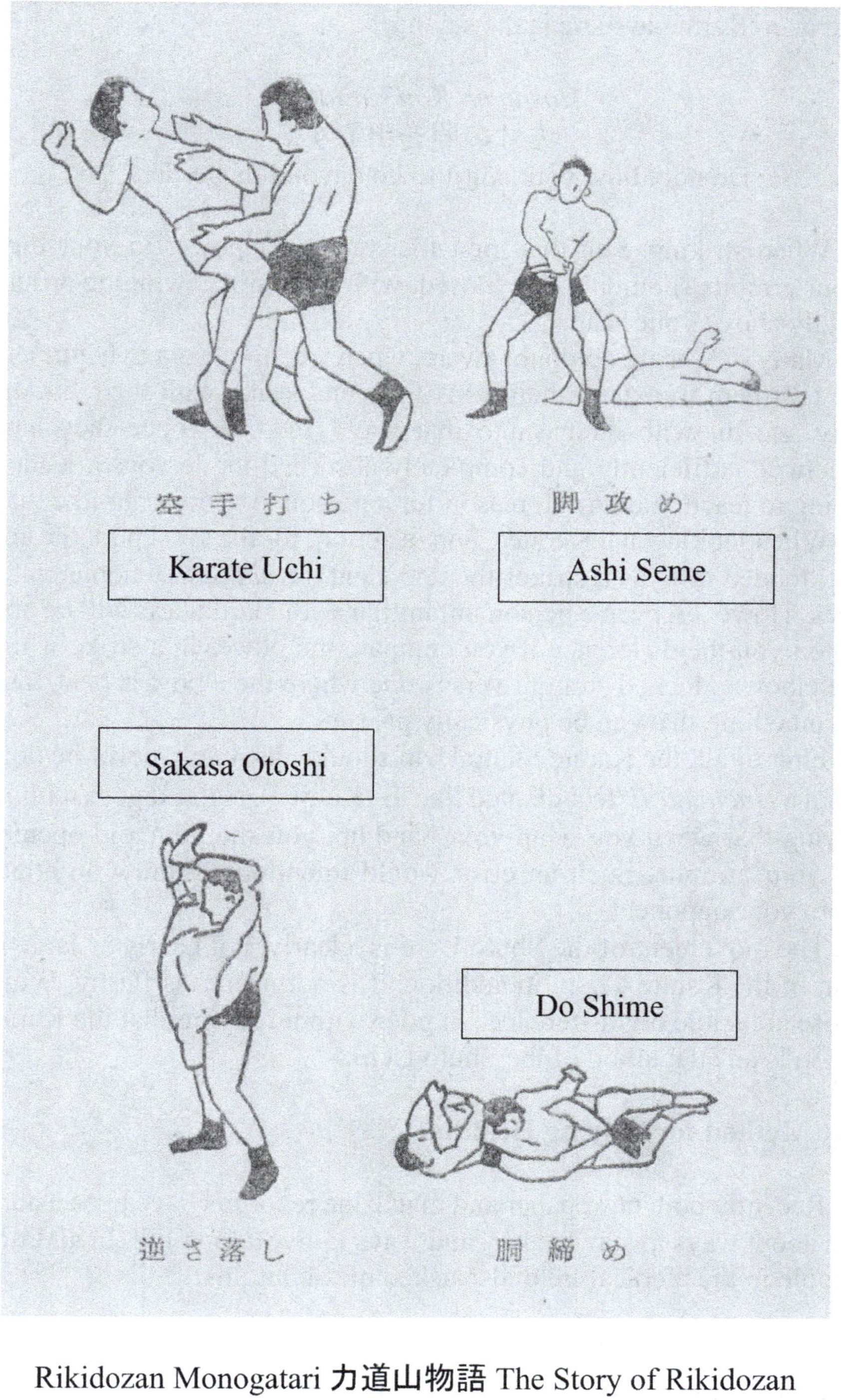

Rikidozan Monogatari 力道山物語 The Story of Rikidozan
Mitsuhash Kazuo 三橋一夫
1954

Just as in Karatedo there is the saying,

Hosun no Kan wo Idezu
方寸の間を出でず
Do not allow your spirit to go beyond its borders[92]

When striking with this, first focus all your power in your hips. Your armpits shouldn't be exposed, with your arms swinging around wildly above your head.

Many of you are no doubt aware, when teaching Kata to beginners, we tell them to extend their arms fully and strike with their fist this way and hit with their Shuto that way. However, once they have practiced sufficiently and completely absorbed the lessons a Kata is trying to teach, there is no reason for a person to be moving like that.

When looking at a Karate Chop in action, for the most part, the arm is extended fully as it strikes the opponent's *Kubi Suji*, tendons in the neck. However, even a person unfamiliar with martial arts will be able to tell you the difference if you compare the power in a strike where the elbow is locked straight versus one where the elbow is bent. This is something that can be physically proven.

First of all, the Karate Shuto Uchi should absolutely never be done from a *Hachi no Ji*, feet planted like the Kanji 八 eight. It goes without saying that when you whip your hand up, you should avoid opening up your armpits. Such an error would intentionally draw an attack from your opponent.

The movement of the Shuto Uchi is, clearly, not nearly as large as that of the Karate Chop, in addition, it is not nearly as flashy. While those are some big differences, it goes without saying that the Karate Chop is an adaptation of the Shuto Uchi.

My Method for Staying Healthy

Recently both newspaper and magazine reporters have been asking me about ways to stay healthy and ways to live a long life. In addition people in the medical field also asked me about this.

---

92 This phrase is more Chinese style than Japanese, however as he was a scholar of Chinese literature it is not clear if it is something Funakoshi Sensei wrote, or a quote from another source.

However I don't have a particular method for staying healthy or a method for extending the life. That being said, as I am nearly ninety years old, clearly I have lived a long life [the truth be told, I feel that I'm just getting started] so I would like to talk about my outlook on daily life and method for staying healthy.

As I said at the beginning of this book I was born early after only seven months and my parents as well as most of my relatives thought I wouldn't live to see my third birthday. Is that probably the main reason that I am still alive today. To put it simply, *Muri wa Shinakatta*, I never overdid it, or went beyond my capacity.

For example, with regards to eating habits, starting from when I was a child up until I reached my present age, I have gotten in the habit of eating hot foods in the summer and cold foods in the winter. Therefore, in summer I don't enjoy ice cream or other frozen treats. Instead, if I want to eat cold foods, I do so in winter.

One more thing I do relates to clothing.

As most of you are no doubt aware, the weather in Okinawa is extremely hot. Even at this age, despite being born in that climate, I don't wear knitted undershirts or knitted long-johns under my pants in winter.

In addition, I don't use *Hibachi* stoves, *Kotatsu* heated tables, or *Yutanpo* hot water bottles. No matter the season, spring, summer, fall or winter, I sleep with a stiff *Hanagoza* mat on top of my Futon and rest my head on a wooden or rattan pillow. I only use one Futon below me and one Futon blanket above. Even in winter I never use a wool blanket, instead relying on the same thin Futon blanket as in summer.

The reason I sleep like this is probably because I lived a poor life from a very young age. Even today I still live in a rented house and my only luxury is insisting on a two story house. Further, I always insist on having my bedroom on the second floor.

The reason I keep my bedroom on the second floor is because all that walking up and down the stairs strengthens my legs and hips. I would have to say this has definitely helped to keep me healthy. Even today I still continue to live in a two story house and it has done wonders for keeping me in shape.

Ichi-ju Issai, rice, miso soup and pickled vegetable
Cooking Style Book 料理のスタイルブック
By Yamashita Shigeru 山下茂 (1898~1989) 1951

One other thing is that even after getting married, and even at my age, I still fold up all my bedding every night and place it in the closet, or, if need be, I hang it outside in the morning and then bring it in again and night at set it up. I have never asked my wife, or a family member to do it for me, much less have a daughter-in-law or grandchild haul my bedding around. Thinking about it now, this little thing may well have contributed greatly to my method of staying healthy.

"Doing everything myself" is probably the main thing I believe in. whether its cleaning my room, drying my books and scrolls to prevent them from getting bugs once or twice a year, I never ask anyone else to help me.

I don't know myself if I am like this because I clean habitually, or because I don't like other people doing things for me, buy whatever the reason it has been a habit of mine for years and relates to why I am healthy.

The one thing I never fail to do every day is take a bath. It goes without saying that people that like Furo, taking a bath, all do this, however when I bathe, I don't particularly enjoy an extremely hot bath. I prefer a room temperature bath and just basically wash all the sweat and grime off my body. I also frequent the Sento, local public bath, so much so that the owner got to know me. Since I was a senior citizen, the owner would pour buckets of water over my back to help me get clean and even thump my shoulders and massage me, but as it kind of tickled, it wasn't to my liking.

I usually stopped him half-way even though he wasn't finished. Further, I never go to a masseuse. Sometimes one of my daughter-in-laws or one of my grandsons or granddaughters would say,

"Granddad you are getting elderly so…"

And offer to start rubbing my shoulders, but as my shoulders don't really get stiff, I feel like I am being treated like an infirm elderly person and have never accepted such offers even once.

I strictly adhere to the *Hara Hachibu*, only eating until eighty percent full. I try to make my meals primarily vegetables, and am a firm believer in *Ichi-ju Issai*, eating a simple meal of rice, miso soup and pickled vegetable. It goes without saying that I don't hate eating different kinds of meat, however I always refrain from eating until I am full. I believe with regards to my method of staying healthy, this is the most important point.

I don't smoke. I also only rarely drink Sake. My revered teachers repeatedly cautioned me about this as it would affect my Karate training, thus I am not in the habit of reaching for my *Sakazuki* Sake cup. As my revered teachers always used to say, "If you are drinking Sake with ten people, or drinking sake with twenty people, or drinking sake with fifty people…you should be considering what you would do if all those people were enemies. Even when drinking Sake, you can't forget about things like this."

In the morning I am quick to rise from my Futon on the floor. I then put all my bedding away, thereby clearing the floor. Next, and this may be difficult for the younger generation to understand, I clean my room, making sure to dust off the framed pictures of the Meiji Emperor wearing his court dress, taken in the final year of his life, which was given to me by my children, along with the picture Saigo Kichinosuke gave me of his grandfather, Saigo Takamori.

After that, I my routine is to go through some Kata using slow, easy motion, wash my face and eat breakfast. It may be due to the fact that I am getting on in years, but I am in the habit of taking a nap in the mid-morning hours. In the afternoon I typically pass the time by reading or drawing calligraphy and I try to use reading glasses as little as possible. However, when I am writing a letter with a pen, or am sitting in front of pre-lined manuscript paper, I must, much to my chagrin, rely on reading glasses.

In addition, my hearing is fine and though I have dentures I am able to eat any sort of food. However, when I am talking with someone, they sometimes come loose, so I continue answering while trying to hold the denture in place, which leads some people to say that I am hard to understand. Further, as I am a bit sensitive about that, I am sometimes not able to answer right away, and I find that rather irritating. However that is not my fault, the fault lies with the denture. It goes without saying that if I get a new set of dentures the problem will go away.

As far as what my work consists of these days, it can be summed up with one word, "Calligraphy."

Many Karate students request calligraphy from me after they graduate University or move to another part of Japan.

I have always enjoyed doing calligraphy, even from a young age. One habit of mine is always grinding my own ink, something that, perhaps due to my personality, can't allow another person to do. I slowly and methodically grind the inkstone in water and that helps get me in the mindset to finally pick up my brush and write.

Since that is how I work, it is difficult to handle all the requests I get, so it is common for a request to sit around for several months before I can complete it. It is unfortunate that some no doubt attribute the delays to my advanced years or lack of vitality, however the real reason is that this is how I work.

I often go out by myself. In my opinion it is not a good idea to be inside the house all day, so, in order to get some exercise I go for walks. Unfortunately, this leads many people to worry about me.

Though I have always been active and energetic, on the other hand I am ninety years old, so it doesn't come as a surprise but sometimes I say to my eldest son's wife, "Tell your husband to be careful when he goes out in Tokyo, there are a lot of automobiles and trains going back and forth at a furious pace."

To which she retorts, somewhat scathingly, "I see, well grandfather, exactly how old are you?"

Which never fails to bring a smile to my face.

I have to admit that if someone who does not know me were to catch a glimpse of me, they would most likely see me as a decrepit old man walking around. From long ago, members of the Samurai class were only allowed to walk using *Suri-ashi*, stepping forward without lifting your feet off the ground. I still walk in that style today but to an observer it would seem like I am dragging my feet across the ground in a shuffle.

Moreover, now that my hair has gone completely white, whenever I am set to attend a university Karate match or other such event, the sponsors want to send a car to drive me over, however I am adamantly opposed as I can easily handle taking the train and changing lines at the appropriate point and getting off at the correct station. Even now I still set off alone to Kamakura, or any station along the Chuo Line, without a family member or guide to assist me.

Sometimes when I am on the way somewhere I will run into a Dojo member. They are invariably astounded and will typically escort me to my destination or home. In their eyes they see a venerable ninety-year-old out by himself and are unable to fathom how such a thing occurred.

That being said, unfortunately my memory is getting a little fuzzy. I will sometimes confuse the Toden "Capital Electric" streetcar station Hacho-bori and Koshimae-bori or get on the right train line that goes to Kamakura but accidentally board the one going in the opposite direction to Sakuragicho.

Other times I will confuse the station names Suido-bashi with Suido-bata but, then again, as young people also sometimes make this mistake, I believe things like this should be looked at as a simple mistake and not proof that I am old.

One other thing that I am having trouble with is the sheer number of students in the school Karate clubs. The numbers have grown dramatically, and I can't remember who is who. First of all, I forget their names. Then I forget what school they go to. I can usually figure out if they are wearing school uniforms with brass buttons embossed with the school crest, but young men will show up to looking resplendent in a suit to announce that they have graduated university and I frequently can't recall who they are or where I taught them.

Occasionally someone I taught thirty or forty years will visit me on a trip Tokyo. It may have been ten or twenty years since I've seen them and I am really in a pinch trying to recall who they are.

The person will invariably remember me quite well however, recalling one person from the tens of thousands I have taught is no easy task. I am often at a loss as to how to respond to their greetings until a member of my household gives me a reminder.

I have no choice but to try and gloss over the fact I don't remember them with a, "My how you've grown!"
When I say this, someone else in the house will comment, "Isn't it a bit rude to say *Look how much you've grown* to someone who has become a successful adult?"

However, I am always thinking back to an earlier era of my life. My life can be summed up as *Karatedo Ichiro*, a life spent travelling down the path of Karate, and now there are successors here and there all over Japan. I can't really explain it but I am filled to overflowing with gratitude.

Nowadays, one thing I enjoy above all others is spending time with the younger members of the Dojo.

Four or five years ago I was invited by a group of Dojo members to Shimoda City in Izu. The group that invited me was very worried I would be exhausted by the train ride to Izu and then the bus ride to the location, however I wasn't tired in the least.

The dozen or so youths that met me at the end of my trip were almost overwhelming in their concern for my well-being, practically carrying me to the Ryokan inn. As I was elderly, the innkeeper had prepared a large room on the first floor, however seeing as the rooms on the second floor had spectacular views, which would be wonderful to wake up to, I insisted, quite strongly, on a room there.

As it turned out the innkeeper, the maids as well as all my Karate students were now even more worried, working themselves to exhaustion escorting me to the bath or to the wash basin. To alleviate some of their worry as I climbed up and down the ladder-like staircase I took each step with a Kiai of *Ya, Ya*. This seemed to alleviate the concerns of the young practitioners. I heard that some of the youth asked about me, “Is your Sensei really over eighty years old?! He seems ten or twenty years younger than that!”
I had to laugh at that.

I must say that I thoroughly enjoyed this that trip. The young people were all thrilled to have me there. The whole time I was there the one thought that welled up within me was, it’s not time for me to die yet.

Though Karatedo had developed and expanded it was hard to say that it had become popular with the general public, so I realized I still had work to do. In order to further develop this art, I have no choice but to go on living.

Regarding the Public Bath

In Buddhism there is the saying,

Sokuze Dojo
即是道場
You can train anywhere, just as you are

Recently some of the younger members of the organization are of the opinion that Karatedo training only happens at the Dojo. They believe that Karate training consists of participating in lessons where head instructor teaches a Kata. However even if such students become a technically proficient Karate practitioner, they have not become a martial artist in the true sense of the word.

Any Karate *Shugyosha*, person on the path of learning Karate, must keep in mind that there are Karate Dojo anywhere and everywhere as you go about your professional and personal life

Typical late-Edo era Sento, public bath

賢愚湊銭湯新話 *Wisdom and Folly Mixed in New Bathhouse Tales* by Santo Kyoden 山東京伝 (1761~1816)

For example, and this may sound a bit strange to people that don't practice Karate, but you can even find Karatedo at the public bath. This is probably something that never enters the mind of younger practitioners.

That being said, I would like to emphasize that *Sokuze Dojo,* you can train anywhere, just as you are, is important not only for those training Karate, but for any person seeking to refine themselves as a human being.

When we encounter someone from the neighborhood, we always exchange greetings like, "Good morning!" or "Good Evening," and if it is a hot day we say, "It is very warm today!" and if it is cold we say, "It is quite chilly today!"

It sort of goes without saying that this is something that all people no doubt do, and I can't imagine any that don't.

However, as this is such an everyday interaction, the fact that such greetings are important can often escape us. While this may seem like my book is taking a reactionary conservative course when these days when everyone is talking about liberalism and democracy, I think people should consider their manners and etiquette when interacting with their neighbors and acquaintances. [93] You should strive to interact with them the same way you treat members of your immediate family, as in your parents, grandparents, brothers and sisters.

It goes without saying that this is obviously something everyone should do. However it is so "obvious" that sometimes it gets forgotten, which is something that should not be happening.

This doesn't just apply to those training Karate, but something every person should consider to be of fundamental importance. In particular I would like to ask those who are on the path of intensive Karate training to focus on this aspect before working on refining your technique. Love and respect the way of the fist, love and respect yourself, in addition to loving and respecting your family and, finally, loving and respecting your neighbors. This will result in you becoming a patriot and loving and respecting your country and countrymen. If you train with love and compassion while using *Seiken*, true fist and being true to the underlying principles of Karate, then have no doubt that as a person you will develop and flourish.

I will give a specific example.

---

[93] Funakoshi Sensei uses the English word "course."

Both at home and at the Sento, community bathhouse, I am sure most people notice the condition of the bucket used to pour water over yourself. You can see whether or not the person who washed before you has left dirty water in the bucket or has left the bucket in an untidy manner.

If you are the next person, and the bucket has been left in an untidy manner, you must first wash out the bucket before you can begin to bathe. If the wash bucket is stacked neatly upside down and ready for the next person in line will naturally want to use that one. That is just human nature.

Obviously, unless that person is completely oblivious, the next person isn't going to want to use a bucket full of dirty warm water, and instead will prefer the bucket that has been neatly placed upside down in the appropriate place.

I recommend trying it next time you are at the public bath. After you have finished washing and before you get in the bath, even if it is just your own bucket, try rinsing the bucket you used and placing it back upside down in its proper spot. A person can wash out one or even two buckets. If you were to do this, the next person to use the bath will have a pleasant and easy time since the buckets will be all arranged neatly. It follows then that the person who succeeds you is also likely to rinse the buckets clean and leave them neat and orderly.

Just by doing this, all the regulars at the local public bath can enter and bathe in a pleasant, stress-free environment.

This applies to when you get into the Yubune, "hot water boat" or large bath. Quite a few people that get in the Yubune holding their Tenugui cloth and rinse it out in the tub. I suppose they don't realize their cloth is dirty.

Washing yourself in the bath no doubt feels good. However doing so means all the hot water becomes dirty. So the next person that gets in the now soiled bath water will also dunk his cloth in the hot water and wash himself. His intention is to no doubt clean himself, however as the previous person had soiled the water with his cloth, he will not know that he remains dirty.

Sometimes a customer wishes to shave, but another bather is sitting in front of the mirror, blithely washing his head and body. So the man has no choice but to try and shave while getting only fleeting glances as the mirror. If someone were to suddenly bump him, a terrible situation could result. This applies to when you get out of the bath as well.

Men using Tenugui cloths to bathe

賢愚湊銭湯新話 *Wisdom and Folly Mixed in New Bathhouse Tales* by Santo Kyoden 山東京伝 (1761~1816)

In addition, it is a simple thing to return the basket you keep your clothes in, even just your own, to its original spot.

Of course there are those who say that is what the hostess is for, however I think we recognize their efforts at keeping the place running well by taking five or six steps back to where the basket belongs. This simple action will no doubt make her job much easier.

I don't remember exactly when, but I learned something when I read *The First Half of my Life* by the founder of Kodansha Publishing, the late Noma Seiji.[94] At the time, as you might expect, I was going to the Sento every day. As I entered the person at the front desk would immediately greet me with, "Welcome!" and as I left would say, "Thank you for coming!" At the time I was pondering whether or not I should be responding to such greetings with a greeting of my own. *The First Half of my Life* discusses how you should be aware of the language you use. I suddenly realized that when the front desk person calls out, "Welcome!" it is the same as the everyday greetings, "Good morning," "Good Evening," "Thank You!" and "Goodbye!"

So after reading that, the very next day in fact, when I slid open the rattly door to the Sento and when the front desk clerk said, "Welcome!" I immediately replied, while smiling, with an enthusiastic, "Good evening!"

At first the clerk looked at me a little puzzled, but then returned my smile. On my way home the clerk said, "Thank you!" and I replied, "Sayonara!"

After that the clerk instantly recognized me and always bowed politely. I responded in the same manner with a bow of my own. I continued to do this from that day on. Then, and I am not sure exactly when this happened, the clerk's greeting switched from a strictly professional style to a warm style of something akin to true friendship.

Thus, though the connection between the local bathhouse and the customer seems to be a transactional one on the surface, it is possible for that to become a warm, personal relationship.

I always tell students in my Dojo that any person who is only concerned with themselves and does not, for a moment, consider whether they causing trouble for others does not have the necessary qualifications to embark on the path of Karate training.

---

[94] Noma Seiji 野間清治 (1878~1938) Noma worked as a junior high school teacher in Shuri in 1905~7, so he may have known Funakoshi Sensei through work.

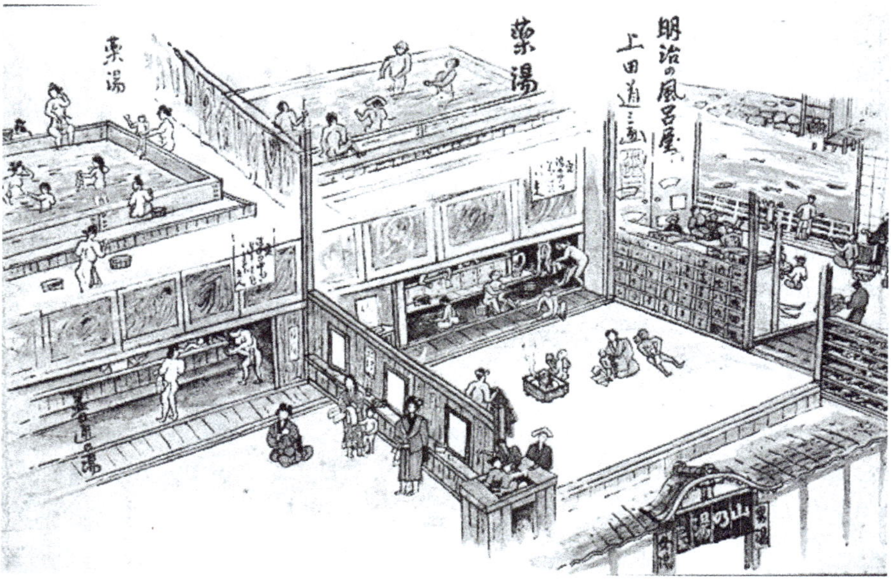

Illustration of a late 19th century public bath. The central partition divides the men and women's sections and a (typically matronly) hostess sits at the front, collecting money and managing customers. There is a dining area to the right and Chinese medicine-infused baths at the top.

On the other hand, I have noticed that people that have continued to train Karate without ever giving up, have included this aspect in their training life. Frankly, *Karate Shugyo*, intensive mental and physical Karate training, is not something that can be done half-heartedly, and people that do so do not continue in this art for long.

This topic now comes up every year around April, when a great number of freshmen apply to join the university Karate clubs. The new students give a variety of reasons for joining, but the most common are to strengthen their body or to improve their mental fortitude. I've heard that some, and while they don't say this openly, join the school's club because they are under the misguided impression that training will make them stronger and they can use Karate as a weapon during fights.

As it turns out, students with this attitude grow discouraged and after only three to six months they stop showing up to Karate club training. Those that seek to train Karate under the misguided goal of using it for fighting are not the type of people that can continue for long.

Only those attracted to exploring the deep meaning and refined ideals Karatedo continue to train. Such practitioners are captivated by the depth of this art, fascinated that it is inexhaustible no matter how much they explore it. That is part of the fascinating mystery of Karatedo.[95]

---

[95] Former Chairman of the Shotokai recalls,

The place at the public bath where Funakoshi Sensei washed himself off gradually became Sensei's "designated spot." He chose the water faucet that was the furthest from the main bath, thus it was close to where the soiled water would drain and collect. Typically, bathers would choose to wash near one of the water faucets closest to the Yubune, large bath. Sensei, on the other hand, purposely eschewed the prime spots instead always using the seat the furthest away.

If the Sento was crowded and someone happened to be using Sensei's preferred spot, he would sit in Seiza at the back and wait, while holding one of the buckets. This habit of an elderly gentleman patiently waiting in Seiza gradually became widely known and respected in the area, even amongst those who did not know he was a renowned Karate expert.

-Motonobu Hironishi (1913 ~ 1999)
1976

Karatedo Rokkun, The Six Precepts of Karatedo

If someone were to ask *What is Karate*, probably the easiest way to make them understand is by explaining Kata. However, that is not the subject of this book, so I will not include an explanation. What I would like to discuss next is something that I often talk to students at my Dojo about: self-discipline and humility for those following this path. This is an important discussion for those wishing to grasp the overall feeling of Karate.

The First Lesson
Karate Shugyo wa Shinken denakereba Naranai
Karate Training Must be Done With Total Commitment

The word *Shinken* means a lot more than its standard definition of "serious" or "diligent." What it means is, and this applies to whether you are standing or sitting, raising your hands or moving your feet, you should always be "Maintaining an awareness of your opponent." For example, if you were to strike with your fist, your attack must be done with the mental commitment that your punch contains every ounce of power your body can generate in order to completely submit your opponent.

You must enter a state where you understand the consequences of, "If this strike misses, I will be defeated."

However, if you devote months and years of your training time only to moving your arms and legs and leaping about, then your training might as well have been in dancing. In such a situation, no matter how many months and years you spend you will not be able to understand the essence of Karate.

Shifting to *Shinken*, fully committed training, is not confined only to those who study Karatedo, any pursuit in the world can greatly benefit from this mindset. It goes without saying that life itself is, in every respect, a *Shinken-Shobu*, duel to the death.

Unless you are joking around, approaching things with a casual attitude of, "Give it a shot, if you mess up, just try again" will not allow you to achieve much of anything in life.

The Second Lesson
Rikutsu wo Nukinishite, Moku-moku to Shugyo ni Shin-shin wo Uchi Komu
Abandon theory and instead focus on training silently and steadily, drilling the essence of this art into both mind and body

This applies to everything. Those who lack dedication are the very ones who tend to argue and make excuses about the theory. For example, a man is taught one Kata. Eventually he says, "No matter how much I practice I can't execute it properly. What should I do?"

I am always surprised to get questions like this as the person has only been practicing the Kata for two or maybe three months.

Consider *Kiba-dachi*, Horse Riding Stance. Though it seems like an extremely straightforward Kata, mastering it requires daily intensive training until your legs feel like wooden Bo. It is not something that can be done after six months or a year of training. Thus saying, "No matter how much I train I can't do it" after only a few short months is rather absurd.

Training is not something done with words but done with your whole body. If you are unable to execute a technique, you should be asking yourself, "If other people are able to do it, then there is no reason I shouldn't be able to. What am I doing wrong and how do I correct this?"

Then you have force yourself to repeat the technique and repeat it again, forge yourself and forge yourself again despite the suffering. This is the true meaning of *Keiko*, martial arts training. People tend to immediately forget the outline of what they are taught. However, if you have thoroughly ingrained a technique into your body, it will stay with you for the rest of your life.

The Third Lesson
Hitotsu no Waza wo Manandara, Tettei surumade Hori-Sagete Manabi Tore
In order to learn a single technique, you must relentlessly dig down until you find the underlying lesson

There are many fundamental techniques in Karatedo and there are many types of Kata. However, the mindset of trying to quickly memorize them all is the wrong way to go about training.

Trying to memorize all the techniques and Kata while not understanding the meaning of each Kata is a mistake. For example, even if you were to memorize every technique, you would hardly be able to demonstrate that you understand the flavor of each.

In the end, this is not the way to train as all the techniques would end up in a jumble and your skills would not be unified. The reason is because you have not devoted time to learning them each separately, thus you do not understand how techniques relate to each other.

In the end, if you do not learn each Kata separately, they will all be scrambled and you will fail to see how they interconnect. This is not the way to learn. You must learn them separately one by one. Trying to learn all the Kata at once means that as soon as you memorize one, you forget another.

On the other hand, if you focus on developing a thorough understanding of one technique, that depth of knowledge will naturally carry over to learning other techniques, which, in turn, will naturally allow you to develop a deep understanding of those.

In conclusion, though there are twenty or so Kata, it should be clear that many elements overlap, so the total number is much smaller than it appears. Thus once you have a total understanding of one Waza, or technique, when you later see a Kata, all you need to do is ask about the outline of it and you will rapidly be able to absorb it.

There is a story about this. It has been written about before and can be found here and there in various books, but I like to include it here. It concerns a very famous *Gidayu*, a man who orates the story in Bunraku puppet plays. This man had a very strict teacher when he was still an apprentice. No matter how many years passed, his teacher would only allow him to recite the tenth chapter of the Taikoki, *The Tale of Toyotomi Hideyoshi*. Day after day, ten times a day, hundreds of times a day he would chant the same chapter. After each recitation his master would refuse to allow him to read any other section, only remonstrating him with, "You aren't ready yet, you aren't ready yet."

Finally, the apprentice lost his patience and made a decision, "I don't think I am cut out to be a reader of *The Tale of Toyotomi Hideyoshi*. I am going move to Edo and find a new profession."

That night he fled his master's house.

Later, while travelling along the Tokaido trail to Edo, he stopped for the night at a postal station in what is now Shizuoka prefecture. It just happed that night, at the hotel, an amateur Gidayu contest was being held. The apprentice of course loved Gidayu storytellers and had studied it for a long time, so he sat off to the side and watched the performances. However soon he was unable to bear it and leapt forward onto the stage as soon as the previous person finished. It goes without saying that the disciple recited the tenth chapter of the *Taikoki, the Tale of Toyotomi Hideyoshi.* As it turns out, when he finished his performance and was preparing to leave, the elderly sponsor of the event approached him saying,

"That was a superb performance, may I ask your name? I presume you are a master performer of some renown?"

The disciple was somewhat embarrassed at the complimentary language and, scratching his head nervously, replied, "No such thing! I am just an amateur who happened, quite by chance, to know the tenth chapter of the *Tale of Toyotomi Hideyoshi* well."

The elderly sponsor was surprised at this and said, "Hard to believe that is true! The way you spoke tonight was only something a narrator at a puppet theater could do! Who on earth did you study under to learn such skill?"

The disciple finally spoke about how his master was extremely strict and his decision to leave. To which the elderly sponsor replied with a strongly worded opinion, "You have completely misinterpreted what your teacher was trying to do. In just a few short years under his tutelage, you have become an expert at that passage. That is not something to despise. You need to turn around and go back, fully apologize to your master and start your training again with renewed purpose!"

The young disciple realized the truth of these words and returned to his former teacher, refined his art as an orator of tales and eventually became a famous performer.

This story may be referring to Koshiji Tayu, but, at any rate, this story has a lot of depth to it.[96] This story can apply not only to Karate training, but to any undertaking in life.

---

[96] Takemoto Koshiji Tayu 竹本越路太夫 (?~1848.)

The Fourth Lesson
Manshin Suruna, Dokuzen ni Ogoruna
Do not be arrogant, do not be self-righteous

Those that seek to present themselves as heroic warriors, walking down the road with arms flexed, forcing others to skirt by the narrow space on either side, may think that people see them as strong, but deep in their hearts those passing by do not respect them. There is nothing more ridiculous than acting tough when you're not really strong.

More often than not, it is beginners who go out of their way to look tough, which results in those same beginners ruining their own standing and respect. They feel that if someone says, "Wow, that guy is studying Karate!" then that serves as proof they are a real practitioner.

The Fifth Lesson
Tan wo Sutete Cho wo Tore
Discard Shortcomings and Adopt Strengths

If you have a chance to watch others practice, you should immediately find some aspect of their training to study. If you see a person who is lacking in some aspect, you should reflect, "Do I have this same problem? How would I correct it?" It should serve as a caution. Humans all have strengths and weaknesses. Therefore you can find something to learn from anyone, further you can also use that as a way to reflect on your own training.

There is an old saying, one that you should chew on well,

If three walk together, one will invariably become my teacher.[97]

---

[97] This quote is from Book 7: Chapter 21 of *The Analects* by Confucius,

The Master said, *When I walk along with two others, they may serve me as my teachers. I will select their good qualities and follow them, their bad qualities and avoid them.*

-Translated by James Legge 1893

The Sixth Lesson
Otagai no Nichijo Koshi Seikatsu Sore Jitai ga Dotoku de Ari, Shugyo de Aru

Our daily lives, both public and private, themselves are lessons in morality and serve as part of our training

While I will try to avoid duplicating what I discussed in the previous section, *Regarding the Public Bath,* I nevertheless wish to add some additional information.

The first thing I want to mention that your *Kohai*, juniors, will develop and grow. This is a fact. There is a tendency to classify students as *Montei*, followers, or perhaps as *Deshi*, disciples, or even *Kohai*, juniors, however I feel this is a dangerous habit. For example, even if it is true they began their Karate journey after you, and are still in the early stages of developing their skill, it is not impossible that you will become complacent and forget the inevitable growth of the following generation. The lesson contained in the fable *The Tortoise and the Hare* is not just for children.

Further, as I always say to the younger members of the Dojo, in the end, your Karatedo must reach the level of faith to be authentic. For example, you need to pray in order to ask for the great compassion and great mercy of the Buddha. I believe that only when we come together to support one another, cultivate the spirit as the highest form of faith, and strive toward the perfection of character, can true enlightenment in Karatedo be attained.

Finally, I would like to talk about something I have heard recently, though I may well be mistaken. Some of the younger practitioners have begun using expressions like *Jitsuroku-gata*, a man of practical ability, or *Sento-gata*, a man who's skills are ideal for the battlefield, or *Jissen-gata*, a man of techniques that have been tested on the battlefield. I believe that by simply looking at the words as they are written, they are clearly childish and are the deceptive words of someone ignorant of the meaning of Karatedo.

People that make such statements are something even beyond what is described in the parable *The Blind Men and the Elephant* and they should be considered to be disparaging the name of Karatedo.

As the Buddhist saint Nichiren said, "While others may read the Lotus Sutra by moving their lips and mouthing the words, they do not read it with their hearts. Even if they read it with their hearts, they fail to read it with their bodies. The ideal is to is to embody the teachings through both the body and mind."[98]

I believe that those seeking to follow the path of Karatedo should fully understand and follow this teaching.

A Curious Tale About the Arrest of a Robber

I believe this incident happened about two or three years after the war ended.

At the time I was around 80 years old, so a lot more active than I am today. I had gone to the house of a traditional poetry master in the Tamagawa part of Tokyo for a poetry reading party. As it turned out this on this particular day the group that gathered was celebrating some sort of anniversary so Sake and accompanying food was freely available, which meant I ended up staying late and catching the last train home. It was still a chaotic time in Japan, and in that era it wasn't safe for a person to walk home alone at night. That being said, I was an old man so unlikely to be much of a target. Thus, I began trudging home from Otsuka Station alone in a light rain.

The area around the station was nothing but scorched earth. Thus, my house, which had escaped the devastation, was visible far off in the distance.

I was wearing an inverness cape and carrying a Western style umbrella as I shuffled along with my aforementioned *Suri-Ashi* walk. The man appeared just as I was passing between Otsuka and Hikawashita. He emerged suddenly, like a black shadow leaping out from behind a burnt telephone pole. He seized the end of my Western style umbrella and said, "Grandpa…."

Hearing this I thought maybe he was someone I knew from the neighborhood and started to doff my cap and bow in greeting. But the man, looking somewhat nervous, said, "Grandpa, can you give me a cigarette?" in a voice that was perhaps intended to be menacing.

"I don't smoke cigarettes." I answered calmly, finally realizing what was going on with this man. I didn't carry a briefcase or even a shopping bag, instead I was in the habit of simply wrapping up all my goods in a Furoshiki cloth.

---

[98] "The Writings of Nichiren Daishonin" Vol.1, P204

That day all I had wrapped up in the Furoshiki were some books and my empty Bento lunch box. I had my umbrella in my left hand and the parcel underneath my left armpit. It was there that my assailant's attention was focused, no doubt imagining I had some money tucked away.

"Grandpa, I don't believe you don't have any smokes on you. I think you know what I'm talking about."

As evidenced by his long-winded threat, the man had obviously not been a thief for long.

"I have never smoked, sorry to put you to such trouble."

"Yeah, well what do you have wrapped up that Furoshiki? Whatever you got, hand it over!"

"I don't have any money or valuables in here."

In response to that, the man fell silent and pretended he wasn't interested in continuing the conversation before suddenly yanking my umbrella and readying to strike me in the side of the face. Looking at his stance, I could see he had many gaps in his defenses.

Suddenly, he swung at me. I naturally rotated my body so I was perpendicular to him and aimed for his groin with my right hand.

"Ah, unnn!"

The man dropped my Western umbrella with a groan. I was squeezing a certain Kyushu, vital point, only found on men.

A patrolman happened to pass nearby so I handed the man over to the police. Thinking back on it, that man was probably just a soldier returning from war who was unable to find a job. He had fallen on hard times and impulsively tried his hand at robbery.

As I have mentioned before, I always tell the students at my Dojo to never make the first strike. Even today, I am still embarrassed about this incident, but it made me keenly aware of the need for greater self-discipline.

Karate and Women

One aspect of Karatedo is the fact that elderly people can practice at an elderly person's level, a strong person can train at their own level just as a weaker person can. Further, women and children can train, each according to their level of strength and endurance.

In addition, you don't need a partner to train Karate. Of course, if you wish to continue to develop your skill, you will need a partner to practice *Kumite*, paired training, and *Jiyu Kumite*, free sparring, however initially you can train alone. In addition, you don't need a particular *Keikogi,* training uniform, nor do you need a Dojo. You can train in your garden or in the living room of your house.

If you are the kind of person that wants to learn different types of Kata then you will have to train at a Dojo to a degree. Frankly if you are just looking to train your mind as well as your physical body to a degree, then Karatedo at that level can be done most anywhere, much like groups that do Radio Calisthenics in a local park.

Recently, for those reasons, there has been an increasing trend for women to join the Karate club at their company, which is a positive trend not only for Karatedo but also for women in general.

Initially, more than a few university students never told anyone else that they trained in Karate, a testament to the fact that public opinion on Karatedo was clouded.

As for who is responsible for that, the responsibility lies with those of us that have been working to expand Karatedo.

The fact remains that, generally speaking, there is a feeling that some things are not suitable for women. Be that as it may, the women and girls who train today find Karate fascinating.

One reason may be Kata. If you look at the elegant movements, they could be seen as the same as dance moves. Recently, I had the chance to see some "women's beauty exercises" on television. Unlike in Karate, these women wear clothes that look like bathing suits and use special equipment. Such things are not needed for Karate so I would like a lot more people to begin training. I often hear, "If a girl starts doing Karate, when she gets married, she will surely dominate her husband!"

While this question may be asked in jest or it may be asked in all seriousness, but the question does come up frequently. I always respond that in no way, absolutely in no way is this true. The question is doubly amusing as most people find that the opposite is true. As evidence to this fact, allow me to point out that Karatedo training begins and ends with a bow of respect, something that women who train have ingrained in their body.

In addition, it is a fact that training makes a woman's face and body appear healthier. Of course, this does not apply just to women, but it is not uncommon for parents to bring a rather unhealthy-looking teenage girl to my humble abode.

After about six months of training, such girls realize they have begun to find the art interesting and, at the same time seem to have left their illness behind.

Thus, I can say with confidence that Karatedo indeed is a "beauty exercise." They are now able to deal with robbers, like the one I mentioned earlier, in addition to defending against men trying to grope them. They can most likely topple such attackers and thereby protect themselves from violence.

I may be repeating myself here, but people that have truly absorbed the lessons of Karatedo, whether man or woman, are not going to be the violent, impulsive person that is a worry to society at large. First of all, people that truly train and live Karatedo are different. If it is a woman, she will be prepared and not go near places where molesters frequent, if it is a man, he will make avoiding places where arguments and fights are liable to break out part of his daily routine.

I frequently talk to the younger members of the Dojo about this, admonishing them, "I am not hoping for you to become strong, rather I hope for you to become weak."

Frequently, they will ask pointedly, "Sensei, I can't understand this at all. We are training because we are weak and want to become strong. If the goal is to become weak, we don't need to train for that, we were already weak!"

I replied to those sharp words, "I am not surprised that you took my words that way, however I would like you all to find the answer and solve that for yourselves. At some point you will all invariably come to a realization and think, *ah, now I understand*."

I have no doubt if those young men continued to train, this lesson would sink in and they would eventually understand what I meant.

Any person who understands how weak they really are, will be able to remain calm and composed in any situation. To embody the epitome of bravery means becoming a true *Yowa-mushi*, weak bug or a coward.

Most Karate practitioners understand that it is essential to repeatedly practice techniques and devote time and passion to training your fists. However, I would like you to remember that you cannot forget that you are also training to learn your own weakness.

Zensin Subete Buki
The Entire Body is a Weapon

When the "weapons of Karate" are mentioned, most people think of the fist, back of the fist, the feet and the legs, however I don't think it would be an exaggeration to say that every part of the body, from the top of the head to the tips of the toes are a weapon. Even counting only the primary methods of striking, there are ten ways just from the wrist to the end of the fingers.

1.	Seiken	True Fist
2.	Uraken	Back fist
3.	Shuken	Hand-fist
4.	Ippon-ken	Single-point fist
5.	Nakadaka Ken	Middle Finger Raised fist
6.	Tettsui	Hammer-fist
7.	Shuto	Sword Hand
8.	Nukite	Spear Hand
9.	Ippon Nukite	One-finger Spear Hand
10.	Nihon Nukite	Two-finger Spear Hand[99]

[99] Illustrations on the following page from *A Karatedo Compilation* 空手道集成 By Keio University Karate Club 1936

Seiken (left) Wakiken (right)	Uraken (top) Alternate Uraken (back of hand)	Hiraken
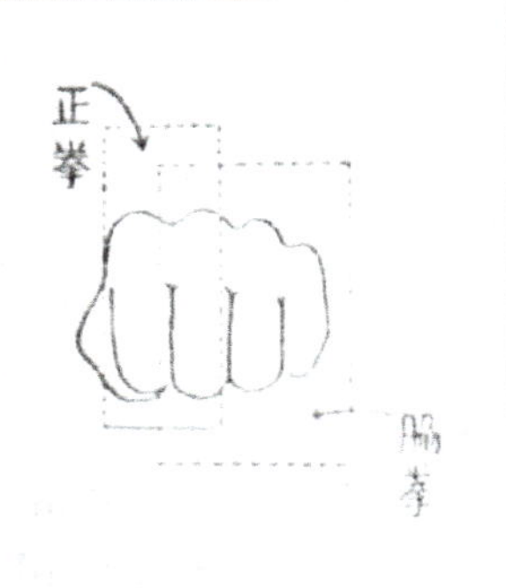	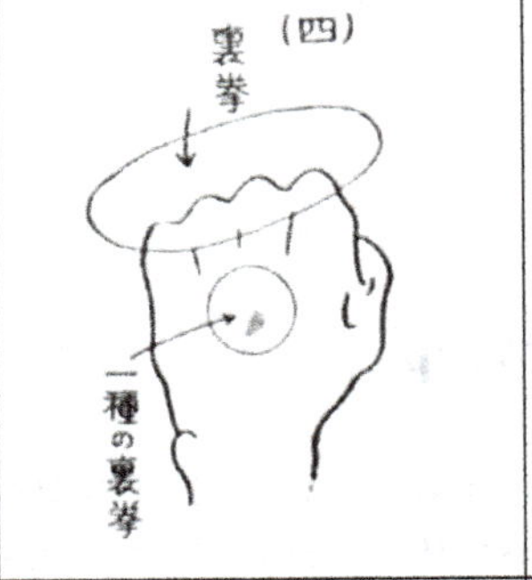	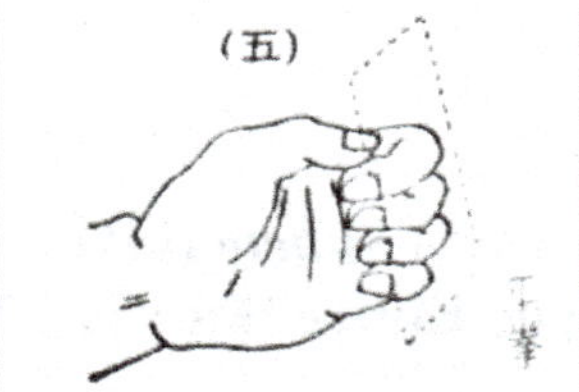
Ippon-ken	Nakadaka Ippon-ken	Tettsui
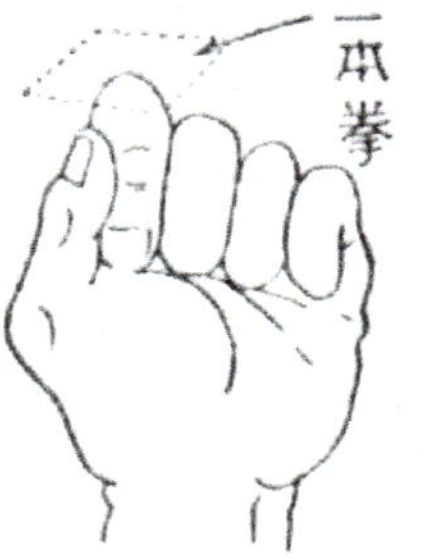		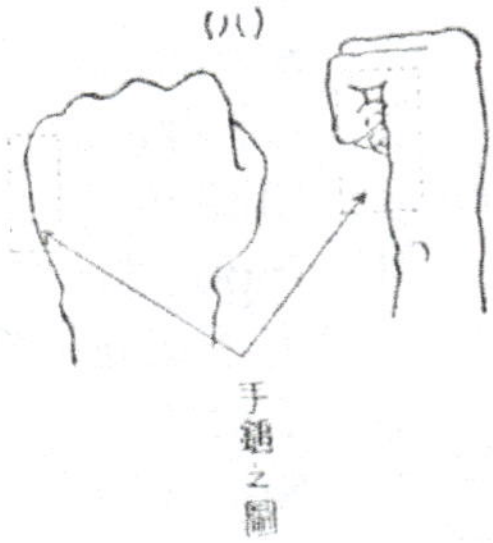
Ura Shuto (top) Shuto (bottom)	Nukite	Ippon Nukite (left) Nihon Nukite (right)
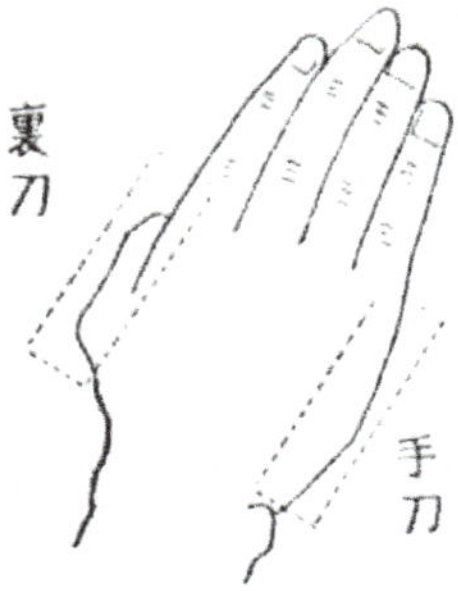	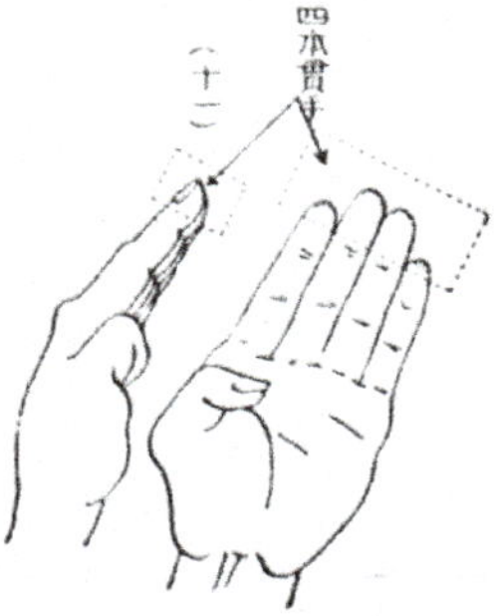	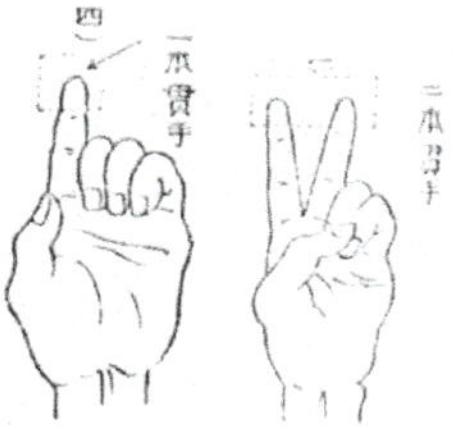

Further, from the ankle on down to the ends of the toes:

Koshi	Ball of the foot
Shusoku	Hand-foot
Sokuto	Sword Foot
Tsumasaki	Tips of the Toes
Enju	Heel
Sokko	Top of the Foot

Other areas frequently used are the wrists, elbows and knees. As you cans see, each and every part of the body can be used as a weapon.

So, I shall next introduce the most commonly used of these weapons as well as how their power is applied. Also, for those that are just training for exercise, I shall also include a simple method to forge these techniques.

First of all, lets separate the different ways of striking into two categories, fist and feet, and start by explaining fist. There are many ways of striking with the fist, but Seiken, True Fist, is probably the most fundamental strike amongst all the fundamental ways of striking. It is also called *Nigiri-ken*, curling your hand into a fist, however this goes beyond simply curling your fingers into a fist. There is a certain way to manipulate each of your fingers. In short, if you curled your fingers into a fist in the normal way, not only would you not be able to put power in it effectively, but you would also likely hurt your own hand when you hit something.

Thus if you were to look at the fist of a Karate practitioner you would likely notice that there was something different in how they curl their fingers into a fist. That is because they make their fist in the following manner.

Illustrations from: *Karate: Ryukyu Kenpo* 唐手 : 琉球拳法 1922		
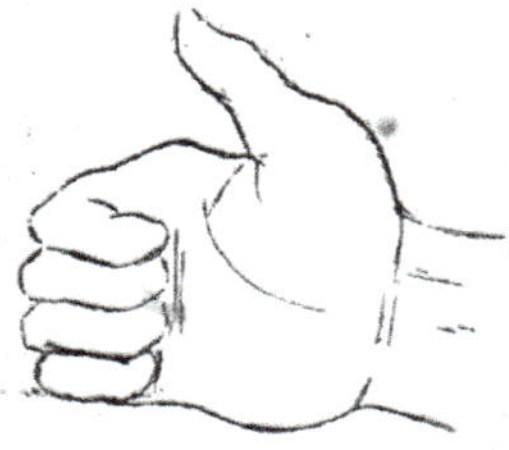		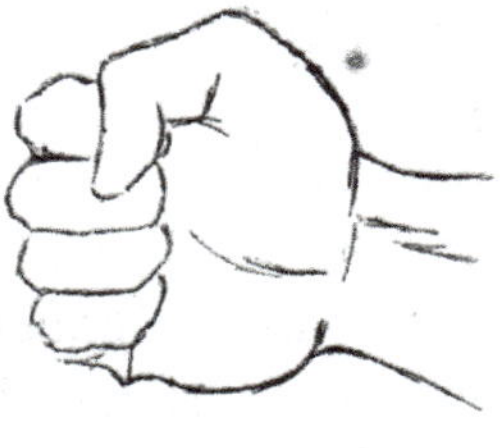
Photographs from *Karatedo: My Lifelong Journey*		
Seiken when viewed from the side	Seiken from the front	Looking at it from the top
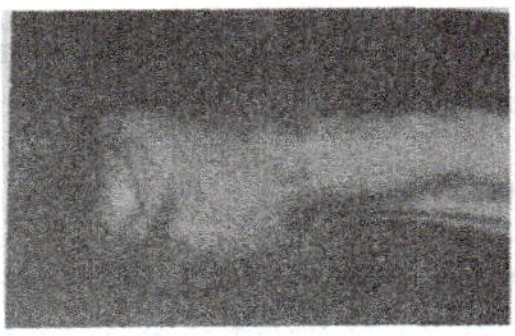	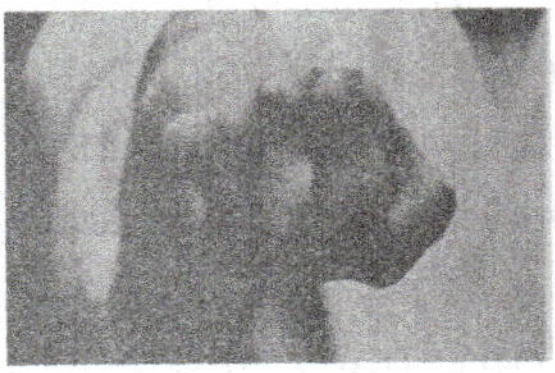	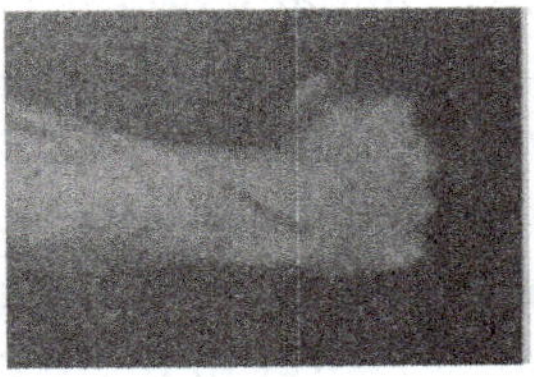

First, bend the fingertips of your four fingers, excluding your thumb, towards your palm as if you are trying to fold them strongly. Then squeeze your fingers in tightly as if you are trying to fold them even further. Finally, wrap your thumb across your index finger and middle finger and press tightly.

The photographs show this. I would like you to pay particular attention to the photo showing the fist from the side. One important point to remember is that rather than being in a right angle, the knuckles of your hand should almost be at an acute angle as you have compressed them so tightly.

It goes without saying that, at least initially, this will be quite difficult and your hands will get tired after only a minute or two. However, as you continue to train you will begin to develop a deeper understanding of it and, eventually, the knuckles of your hands will develop round calluses and your fist will form a right angle. Then you will find that you will naturally be able to form a Seiken fist without thinking about it.

One good way to train this is in the bath. Cover your hands in soap and practice gripping and releasing your fist.

Leaving the thumb on the outside like this has been developed after long years of Karate training. Some people may puzzle over this thinking that it will result in injury to the thumb however, that is certainly not the case.[100] In all likelihood, if you were to strike an opponent with your thumb tucked into your fingers, you would surely injure your thumb.

For beginners, it can be difficult to curl your fingers into a fist as shown in the picture and then strike. There is a tendency for them to allow their wrist to bend so their fist can easily end up turning upward. If you strike like this then the fundamental part of the fist, the Kento, "head of the fist" (referring to the knuckles, or more specifically the first knuckles)[101] will not make contact. Instead, the second knuckles with make contact painfully resulting in a sprained wrist.

Clearly if you strike your opponent with your fist in a Seiken like the one shown in the photographs, then all the power in your arm will be focused directly into the first knuckles of your index and middle fingers.

Seiken can probably best be described as the most vital aspect of Karate. Forging this way of striking requires that a practitioner be motivated to train day in and day out. If you are not able to generate

---

[100] Several Judo and Jujutsu books include illustrations showing the thumb tucked in. The illustration on the left is from an unknown 19th century Jujutsu school scroll. The illustration on the right is from *New Judo Instructor's Guide* By Sugawara Sadamoto. 1926.

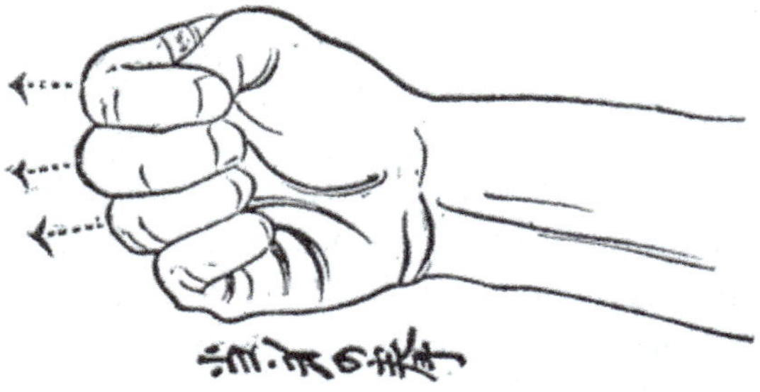

[101] Brackets are by Funakoshi Sensei.

sufficient power and intensity in your punch, then not only Kata but also Kumite will fail to have any meaning.

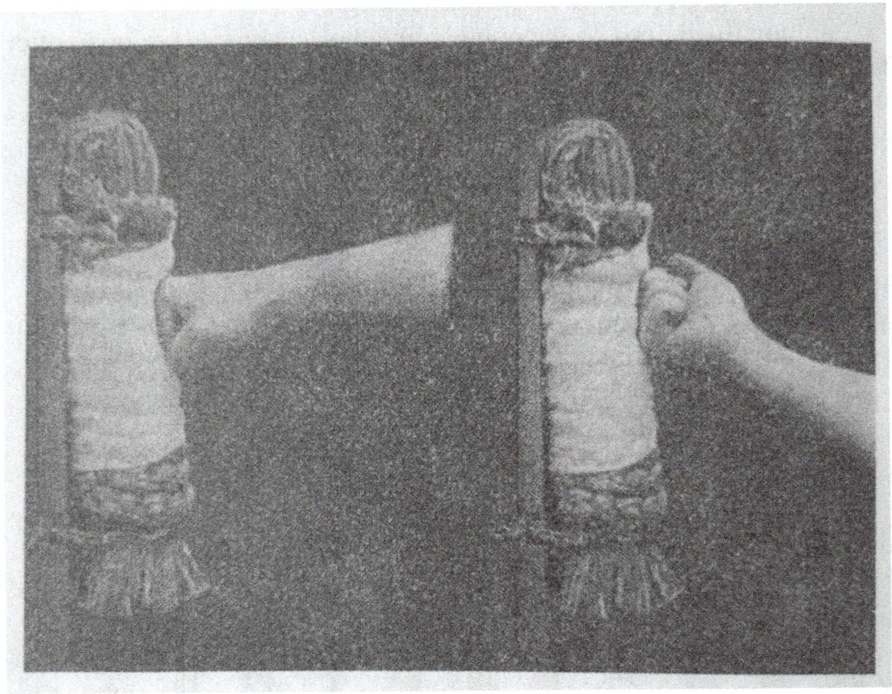

Striking Makiwara with Seiken (left) and the moment you pull back
Photographs from *Karatedo: My Lifelong Journey*

Frankly speaking, everyone struggles to develop their Seiken, however the most commonly used training tool is the Makiwara.

Originally, the Makiwara was not just for forging the fist, but also training the Shuto, the elbows and feet. It would not be an exaggeration to say that powerful weapons are born from Makiwara training.

To make a Makiwara training pole, first prepare a *Tsuki Gashira*, Striking Pole. It can be made out of any material, but typically something with a little flex. It should be about 7 Shaku, 7 feet, long and about 3 Sun, 3 inches, wide. It should be about 4 Bu, 0.5 inches, thick at the top and about 2 Sun, 2 inches, thick at the bottom. Find an appropriate spot and bury the pole in the ground with the thicker end at the bottom, so it gradually tapers up. Generally speaking, about 4

Shaku, 4 feet, should be above ground, however, adjust it so that the head of the Makiwara target is at your shoulder level.

To make the rice straw wrapping, first pound the straw flat to soften it and separate it into three bundles. Make bundles about 2~3 inches in diameter and then tie those to the pole with rice straw rope. Note that if you are a beginner you may find that your hands hurt and the skin on your fists splits, so it is a good idea to wrap a Tenugui cloth around the rice straw. With that, you have finished your Makiwara training pole.

So then, next is how to forge your Seiken, True Fist. First extend your arm and find a position where you can easily reach the Makiwara from *Hanmi*, half-body stance. Next, make sure your knees are bent sufficiently and your hips are lowered as much as you comfortably can. Stand with your forward fist about 5~6 Sun, 5~6 inches, from your knee while you hold your rear fist against your side with the back of your hand facing down. Pull your chin in while keeping your eyes locked fiercely on the Makiwara. Allow your shoulders to relax and droop while you focus all your energy in Tanden, the center of your power.

When doing this the way your feet are positioned is critical. You must be standing solidly as if you are rooted to the ground and anything less than that is incorrect.

This is the ready position for striking Makiwara. When doing this training, ensure the Makiwara is no higher than chest level. If it is too high, when you strike your legs will extend, causing your hips to rise, meaning your strike will lose power.

So then, the way to Tsuki is as follows. Twist your hips so your fist held by your side moves forward, as if you are trying to strike the Makiwara with your shoulder, build up all your power…and launch your Tsuki. At the same time, pull the fist that was forward above your knee back until it is tightly against your side.

Just before your fist makes contact with the Makiwara, rotate the back of your hand upward so that when you strike it drives in with the feeling of twisting. At the same time, ensure that the fist you are pulling back to your side rotates so that the top of your hand is facing down.

Striking Makiwara
*Karate: Ryukyu Kenpo*唐手：琉球拳法 1922

Of course, this rotating of the body will initially prove quite difficult to get the hang of, so I recommend repeatedly practicing that movement without striking before moving on to actually hitting the Makiwara.

As for people that are just using Karate as a method of calisthenics, there is no need to strengthen your fist, so you can just practice repeating the movements as exercise.

The upper part of the board you are striking is thin so it has some flex to it, so if you do not strike it with sufficient force, it will rebound back at you. Thus, the proper way to strike is to punch deep, extending your punch before stopping.

In addition, people that are right-handed should practice with their left hand more, in order to prevent a tendency to prefer that side when doing techniques. The more times you practice this strike, the better. That being said, initially your fist won't be tough enough, so you should decrease the power of your punch in order to practice more repetitions. Once you have gotten used to this drill, throw all your power into it. There is one phrase you should constantly repeat to yourself as you train,

Ryo Yori Shitsu de Yuku
Quality over Quantity

The important points to remember when training Makiwara are twisting your hips while keeping your center of gravity low and focusing all the power in your body in *Kento*, the head of your fist. That being said, anyone who brags about having blisters or calluses on their hands not ever develop a truly strong technique.

Even if you beat the rice straw before making a Makiwara, beginners will still find that their fists hurt. In addition, as beginners are distracted by pain while also trying twist their hips to focus on striking a point 3 Shaku, 3 feet, away, it is not uncommon for them to miss the Makiwara target entirely.

Thus, beginners end up stripping the skin off the knuckles. If your fists end up swelling from Makiwara training, you can treat the swelling and pain with cold water, however if your skin has split, you should take a break from Makiwara training for a week or two.

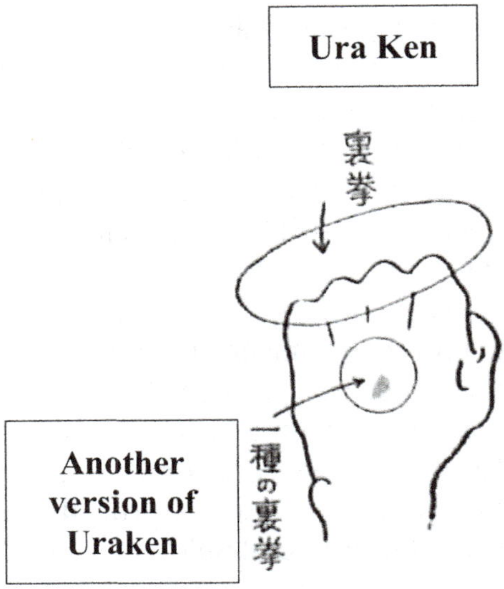

A Karatedo Compilation 空手道集成 1936

Next, I would like to discuss strikes other than Seiken.

First of all, there is Uraken, Back Fist. To make this fist, curl your fingers the same way you did for Seiken, however as you use the back of the fist it is called Uraken, Back Fist.

When using this fist, you are primarily striking with the area around the first knuckles of the index and middle fingers, though when forging Uraken with Makiwara you will be striking with the whole back of the hand.

The Uraken is extremely effective. It is used primarily when striking the face, under the armpit and on the side, however it is most effective against opponents who are perpendicular to you or an opponent facing you that rotates perpendicular.

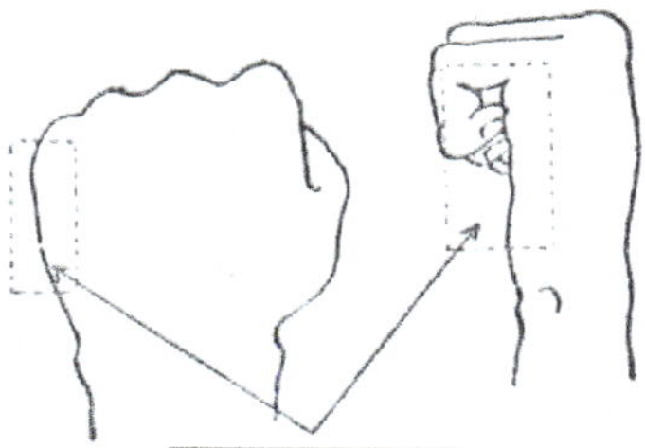

Tessui

A Karatedo Compilation 空手道集成 1936

Next is Tessui, Hammer Fist. Make this fist the same way you make Seiken. You are using the side of your hand by your little finger, in other words what you use to pound on a person's shoulders when helping them to relieve tension. People that are not familiar with this strike probably think that it is not effective however this is completely wrong. If you have forged your Hammer Fist on Makiwara then it can become a spectacularly effective weapon.

Since the side of your fist is soft and flexible, no matter how hard you hit with the Hammer Fist, it doesn't hurt. That is why this weapon is used to hit and sweep aside your opponent's wrist or attack his joints.

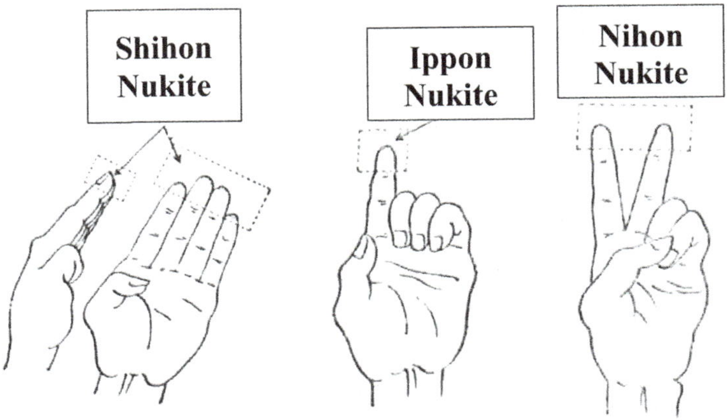

A Karatedo Compilation 空手道集成 1936

Photograph from *Karatedo: My Lifelong Journey*

Nukite, Piercing Hand or Spear Hand, describes using your fingertips to pierce through something.

If you bring all four fingers together it is called a Shihon Nukite, Four Finger Piercing Strike, or sometimes simply Nukite. If you use two fingers to strike then it is called Nihon Nukite, Two Finger Piercing Strike. If you use one finger to strike it is called Ippon Nukite, One Finger Piercing Strike. The photograph above shows Shihon Nukite, Four Finger Spear Hand. Note that your thumb is bent while your four fingers extend straight out. You thrust forward with your fingers aligned and pressed tightly together.

People who are unfamiliar with this strike may think attacking this way will lead to one or more of your fingers getting sprained, however the opposite is the case. If you have trained and forged the ends of your fingers strikes to your opponent's face or Mizo-ochi, solar plexus, can be extremely effective.

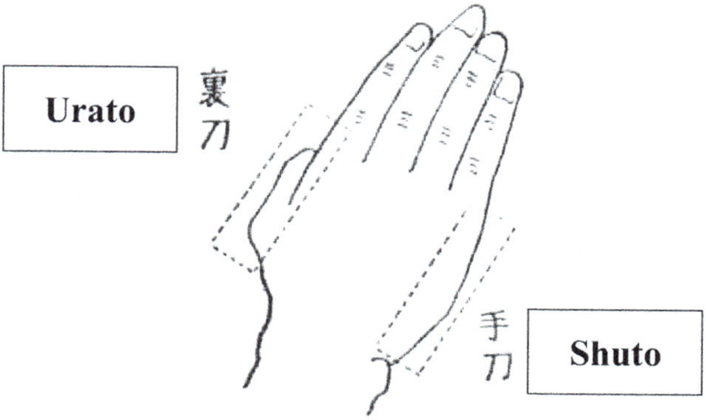

A Karatedo Compilation 空手道集成 1936

The Shuto, Sword Hand, which I explained somewhat in the section titled *The Karate Chop* can be applied in a wide variety of situations.

Your hand forms a shape exactly like Shihon Nukite and you use the side of your palm near the little finger. In other words, if you curl that same hand into a fist, it will be the same spot you hit with Tessui, Hammer Fist. With your hand open, it becomes a Shuto. It can be used to attack the neck, side, arms and legs. To train and forge this part of the body, use Makiwara.

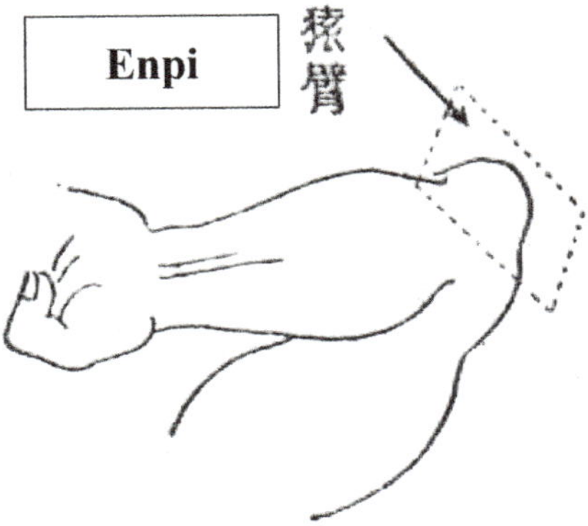

A Karatedo Compilation 空手道集成 1936

Enpi, monkey elbow, refers to the elbow.[102] Karate frequently employs the elbows effectively. They can be used both offensively and defensively. In addition, as the bones in the elbow are tough, with sufficient training even women and girls can take down a large man with one strike. Just as the word *Hiji Teppo*, Elbow Rifle, indicates, it is a powerful weapon.

There are many ways to use Enpi. If an enemy seizes you from behind you can attack him with Ushiro Enpi, Back Elbow. An effective attack on an opponent to your side is Yoko Enpi, Side Elbow, or using Tate Enpi, Vertical Elbow, when faced with an opponent in front of you. Also if an opponent tries to angle up from below you can stop him with Shimo Enpi, Descending Elbow. There is also Mawashi Enpi, Rotating Elbow, for striking after grabbing an enemy and pulling him close.

In addition, your opponent cannot predict where you will strike, in the face, chest, side, head or back.

Further, elbows are not just for attacking, they can be used to defend the chest and sides. If you are thrown down by your enemy and can't use Seiken, then you can use your elbows against the tops of your opponent's feet. One strike from an elbow contains the power to cause an enemy to surrender.

---

[102] "Monkey Elbow" originally referred to any person with long arms, like a monkey. The first use of this term was by the Japanese Confucian scholar Miyoshi Kiyotsura 三善清行 (847 ~918)
If it is thus, people will stand with Enpi (long arms,) shoulder to shoulder at the gate, and even the barking of dogs will cease, with no need for alarm throughout the land.

Koshi

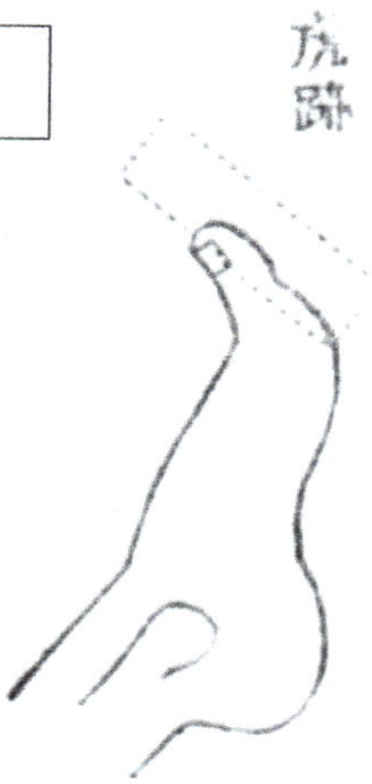

A Karatedo Compilation 空手道集成 1936

There are many ways to use the feet and legs. As the saying goes,

Karate no Tokucho no Hitotsu wa Ashi wo Tsukau Koto de Aru
One particular trait of Karate is the use of the legs and feet

For the most part the attacks are kicks, however sometimes Harau, Sweeps, are used. While your legs and feet are thicker and more powerful than your hands, you will initially have difficulty using them freely until you learn to employ them effectively.

The problem is if you employ your legs poorly, you will end up supporting your entire body on one leg, which will adversely affect your balance. This will mean your movement will be blunted and your leg may well be caught by your opponent. However, if you have developed your skill in this, then you can employ this weapon in order to catch your enemy by surprise.

The name for the ball of the foot *Koshi*, Tiger 虎 Footprint 跡, is a bit difficult to understand just by looking at the Kanji. With the toes of your foot bent back, the ball of the foot will be able to strike powerfully exactly into the depression that forms your opponent's solar plexus.

This is used with front kicks, which should be considered a fundamental strike, just like the Seiken is with the fist. Note that if you failing to extend your toes upward sufficiently and not putting enough power in your ankle will result in injury.

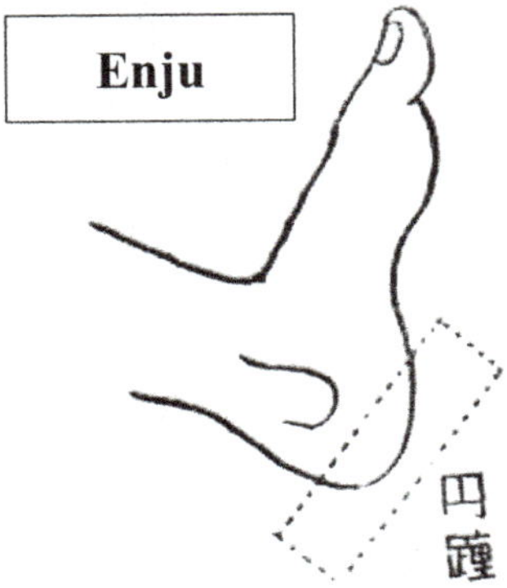

A Karatedo Compilation 空手道集成 1936

Enju, Round Heel, refers to the back of the heel. It is used when kicking an opponent behind you with a rearward kick, or it can be employed when your arms have been seized or someone had leapt in from behind and grabbed you.

Okinawan Wrestling

If I told you that Okinawa does indeed have wrestling, I imagine some readers would be surprised.

As it turns out, the origin of Okinawan wrestling resembles that of Karate with no documents detailing its origin or development. In all likelihood the skills found in Okinawan wrestling are in some way related to Karate, however there is no clear evidence of a connection.

Okinawan wrestling is known as Tegumi. Of course this is an Okinawan word. If I were to write it out in Kanji, I would probably use 手 Hand and 組 Paired. Interestingly, the two Kanji in this word are the reverse of Kumite 組手 paired training, in Karatedo. In my mind, this indicates some sort or connection between Okinawan wrestling and Karate.

When we were young, my friends and I would always have friendly bouts of Tegumi. There weren't a lot of troublesome rules, just a few *Kin-te*, techniques that were banned. Basically, there were four banned techniques:

- No hitting with your fists or kicking with your feet
- No striking with Shuto or Enpi

- No grabbing the hair
- No pinching or gripping the skin

As the rule above states Tsuki was not allowed. This included not only a total ban on punching but also hitting with Karate style Uraken or Nukite.

Tegumi wrestling isn't done in a special ring. A Tegumi wrestling match can be held anywhere, in someone's house or out in a field. Further, unlike with pro-wrestling, combatants don't duel wearing just a pair of shorts. Everybody wears their everyday Kimono. It goes without saying that whenever we wrestled in someone's house the Shoji paper screens got ripped and the Tatami mats got scuffed, which usually resulted in the homeowner scolding us. Thus, we usually wrestled out in a field. There is an Okinawan folk song that goes,

Rikyu Ojarunara Waraji Haite Ojare Rikyu wa Oishi Koishi Hara[103]
Powerful braves, if you are going to wrestle, tie on your Waraji sandal, for wrestling clean up the stones on the field big and small!

As you can see, the song talks about how there are many rocks about and before the match can start, everyone has to work together to prepare the site by getting rid of the rocks and picking up any trash. The bouts begin much like Sumo or wrestling matches do, with both combatants pushing and shoving each other. Then, through takedowns or throws, one person is beaten. The loser surrenders by lightly tapping his opponent to signal his defeat. This is how the matches are decided.

There are many different techniques, however there is one called *Ebi-Gatame*, Shrimp Lock, that seems much like the Leg block and three-quarter Nelson used in professional wrestling today. Whenever I watch professional wrestling on television, I am frequently struck with similarities to Okinawan Tegumi wrestling. Generally speaking, while the winner and loser were decided by the two combatants, this doesn't mean there wasn't a referee.

This person, who could perhaps best be described as an overseer or referee, couldn't just daydream while a match went on as someone could stop breathing or break a bone. When dealing with youths

---

[103] This seems to be a mix of Japanese with some Okinawan words, or Japanese words with Okinawan accent.

determined to win at any physical cost and loathe to admit they are beaten, it is the judge's job to determine when a winner has emerged.

Interestingly, much like a tag-team match in pro-wrestling, where one person is attacked by two opponents, there were Tegumi matches where one person took on two opponents, or one person took on five opponents. Generally speaking, you would be taking on people much younger or weaker than yourself. This kind of bout begins with one person lying flat on the ground like the Kanji *Dai* 大 meaning big, but also resembling a person with their arms and legs outspread. Next, several people take hold of each arm and leg. Then the bout starts.

Frankly, this less of a martial arts bout and more of a game, however those involved treat it like a win or lose match.

Right after I had decided to start training Karate, I would often gather four or five local youths and have a bout. I was interested in building power in my arms, legs stomach and hips, thereby improving my Karate.

I can't say for certain if it helped me in any way physically, but it certainly helped me develop my mental strength.

For example, even if I can shove one opponent 1 Ken, six feet, away, it is not easy to then push a second or third opponent away. Also, as one opponent is attacking you, another can surprise you from the opposite direction. Thus, when faced with multiple opponents, you must learn how to maintain your awareness in all directions. You can learn a lot from this type of training.

Recently, I've heard that back in my homeland, children are playing Tegumi. I presume that, much like long ago, they are clearing the area of rocks and other rubbish before starting. However, I've heard that back in Okinawa there are still relics from the Pacific War like mortar shells and other fragments. I hope that no children practicing Tegumi are injured by such relics.

Foreigners and Karate

Following the war, no matter who you asked, those running Karate Dojo all said that there were many foreign visitors.

Before the war, any visiting foreigner would have been a physical fitness teacher or a reporter. There weren't any foreign people coming to the Dojo requesting to train or looking for answers to questions about Karate. However, following the war, for some reason a great many people, mostly members of the US armed forces, began expressing interest in training Karate.

However, as most are aware, Karatedo and Kendo and other such martial arts were all banned in the years immediately following the war. Apparently, General Headquarters became aware of the phrase,

Ichi Geki Yoku Teki wo Taosu
One blow will flatten an enemy

However, if we called Karate a "sport" then the ban would be lifted, so I got together with other people involved in this art and formed the *Karate Kyokai*, Karate Association, and Karatedo as a sport saw the light of day.[104]

I think this next story was from right around that time.

The vice president of *Reader's Digest* and general manager of the far east division Dennis McEvoy had travelled to Japan and invited me to meet him.[105] The late Suzuki Bunshiro served as my guide and we had a lunch meeting at the Imperial Hotel, which, at the time, was very strict regarding Japanese visitors.[106] I was quite impressed and flattered by the room we were lead to, which was Japanese style with a golden Byobu folding screens and Ikebana flower displays.

McEvoy and I talked for about an hour, and he questioned me closely, at one point asking my age. When he heard how old I was, he laughed and quipped that Karatedo could perhaps be promoted as a method for long life rather than a martial art.

In addition, I believe it was a short time after the ban was lifted, that I was invited to the Tachikawa Airforce base. There I taught Karate to physical education instructors in the Far East Command who were stationed in Japan.[107]

Nowadays, there are a great many foreigners training at the Karate Kyokai Dojo, sometimes there are also women who come to train.

---

[104] Funakoshi Sensei uses the English word "sport."

[105] Senior Editor of Reader's Digest (1895~1958)

[106] Suzuki Bunshiro 鈴木文史朗 (1890~1951) Japanese journalist, newspaper reporter, critic and politician.

[107] Tachikawa Airfield 立川飛行場 in western Tokyo. Built in 1922.

Arawashi Monument beside Kisarazu Shrine
Showa 16 (1941)

Kisarazu Shrine
Showa 13 (1939)

Thinking back on it, the one experience that filled me with the most emotion was when I was invited to the Kisarazu Air Base in Kisarazu City, Chiba Prefecture. By pre-arranged request I presented a completely Japanese-style martial arts demonstration. The base commander, despite it being well outside his working hours, wore perfect military dress, and treated me with upmost courtesy not only during the demonstration but when he picked me up and dropped me off at the conclusion.

Just as you enter the gate of this base, on the right hand side there is a command center and staff office. If you continue on, at the end of the road there is a small garden with a life size bronze statue and a small shrine. The bronze was of a youth standing powerfully pointing out to the Pacific ocean with a flier's uniform around his waist wearing a fliers hat and an eagle with its wings spread at his feet.

It was clearly built when the Japanese Air Force was flourishing and was a shrine dedicated to the spirits of all those who died during training and were thus considered casualties of the war. The Torii gate and the shrine itself had been kept clean and tidy. According to the commander, though Japan was defeated in war, the Americans still respected the spirit of loyalty held by those pilots who gave their lives in battle. Therefore, this shrine and statue had been properly cared for. He concluded by speculating that there may well be the spirits of some students of my Dojo also enshrined here. That was the overall subject of what he said.

I felt overwhelmed with emotion. As I bowed my head towards the shrine, the commander also saluted, demonstrating his respect towards this sacred spot. At that moment, I felt the commander was the epitome of knightliness and as I bade farewell to him, I had a feeling as if a great many worries has been swept away, leaving me spiritually invigorated.

Shortly after the peace treaty was signed a high-ranking air force official invited my organization to demonstrate Karate in mainland America. Obata Isao (graduate of Keio University,) Kamata Toshio (graduate of Waseda University) and Nakayama Masatoshi (of Takushoku University) traveled to various bases in the US over a period of three months to instruct. The official was very serious about this endeavor even making an official plane available to transport the Japanese Karate instructors between bases.

Previously, my organization had been invited to teach small groups of maybe five or six people. This was the first time my organization had been called upon to teach large units of soldiers and I can't express in words how overwhelmed with emotion I was.

I am looking forward to seeing how the seeds spread, not just over the American continent, but Europe as well, will grow and flourish.

Even as I was writing this draft, I received numerous letters from devotees of Karate, asking if I couldn't dispatch an instructor to France, Italy, the Czech Republic, India and other countries. Frankly, I am bewildered as to how all these people ever heard about Karate. As soon as this book, *Karatedo Ichiro*, *Karatedo: My Lifelong Journey* is released into the world, I will no doubt have a lot of work associated with its publication however, needless to say I am very much looking forward to that.

Recollections of My Homeland Okinawa

I am a citizen of Okinawa Prefecture. I was citizen of Okinawa Prefecture before I moved to mainland Japan and I am now. There are some people that believe that Okinawa is not Japan, however they are completely incorrect.

Further, there are people that think Karatedo is somehow a Chinese martial art, but this is also a complete fabrication as it is clearly a *Kobudo*, classical martial art, from the nation of Japan. That being said, there are no historical documents that can offer proof of its origins.

Okinawa being part of Japan can be proven through credible historical as well as ethnological evidence and, even today, folk traditions continue unadulterated from the Yamato period (300~550 AD) and Genroku period (1688~1704.)

In addition, the ancient history researchers that travel to Okinawa report on this. Even today it is common knowledge that there are numerous academic survey teams dispatched to Okinawa.

I would like to give one or two examples of this.

There are many who say things like, *Since Okinawa was once an independent country, it is not Japan*. However this is just a surface-level interpretation and not at all correct. *I see, but if that were simply a surface-level interpretation, how do you explain the fact it was ruled by kings?* However that was no more than a *Noren* curtain disguising Satsuma Domain's foreign commerce.

As everyone is no doubt aware, at the time, the official policy of the Tokugawa Shogunate was *Sakoku*, "chaining off" the country to foreigners, and thus conducting trade with foreign countries was illegal, therefore Satsuma domain could not directly negotiate commercial deals with foreign nations.

Thus, at least on the surface, Okinawa was an independent kingdom, conducting trade with China, when in fact, this was all done under the direction of Satsuma Domain.

In addition, there was a long tradition of Okinawan missions to Tang China, however these were primarily trade missions and served as de facto trade routes.

This has all been thoroughly researched by people in various fields, and is available to read in various publications, preeminent of which is *The Thousand Year History of Okinawa.*

However, as Okinawa was quite removed from the rest of Japan, it took a long time to absorb the culture from the mainland and thus there were many old-fashioned customs that remained. That is why, barring researchers and people with a particular interest in Okinawa, many mistakenly conclude that Okinawa is not Japan based on their visual impression.

If I were to give an example of this, I would point out how the dialect in Okinawa still contains many words dating from the Genroku Era (1688~1704.) Moreover, the late professor Kawakami Hajime of Kyoto University was sent to Okinawa and spent several months in Shuri City in order to research the state of Okinawan society.[108] While there the professor found a place where a lot of elderly people lived and began researching the Okinawan way of speaking. As it turned out the profoundly knowledgeable professor of Japanese literature discovered that after just ten days of research he was able to speak fluently with Okinawan elders.

At a stroke Professor Kawakami determined that these elders were indeed Japanese and, at the same time, was able to prove people living in Okinawa were Japanese.

However, in the following war, the people of Okinawa fought fiercely. Over 170,000 Okinawans died in battle during the war, and the fact that children also battled fiercely as evidenced by groups such as the *Himeyuri Butai*, Lily Princesses Student Corps, *Kenji Butai*, "Healthy Youth Division" also known as the Blood and Iron Loyalist Student Corps, is hardly something I need to discuss in detail.

---

[108] Kawakami Hajime 河上肇 (1879 ~1946) was a Japanese professor and economist.

I am nothing more than one man seeking to introduce Karatedo. It goes without saying that I dearly love my homeland of Okinawa, that gave birth to Karate and where I too was born. In the final years of the war, when defeat was certain, I was trapped on mainland Japan along with Karate.

Obviously today, Karatedo is firmly rooted in every part of Japan. In a sense, that would seem to indicate that my work is over. However, the work to make Karatedo the national sport of Japan and expand it all over the earth begins now. To me, it seems as if I am standing on that starting line right now.[109]

Whenever I think of this I once again recall the bones of my Senpai as well as my revered teachers Asato and Itosu that are interred in my homeland.

When I think back on my *Karatedo Ichiro, Karatedo: My Lifelong Journey,* I realize it began in Japan's better days, and now I wait for my fate in a corner of my war-defeated homeland. It sometimes seems like something out of a dream, though other times not like a dream at all. However the sound that always brings me back to reality, and this invariably draws a smile from me, is the sound of the vital and enthusiastic Kiai shouts ringing out fiercely from the throats of Karatedo students all over the country.

I wonder whether the spirits of my revered teachers, my Senpai, friends and my wife, all of whom were devoted to me and gave me unwavering encouragement and guidance, are listening to me now…

---

[109] Funakoshi Sensei uses the English word "start line."

Collection of Poems and Articles Celebrating Funakoshi Gichin Sensei's Sixtieth Birthday

By Fellow Students

1933

(excerpts)

Sixtieth Birthday Poem

A form and spirit as refined as the peak of Mount Fuji
For those of us who have embarked on this path
His is a name to be cherished and person that conducts himself with restraint
A halberd that has stopped, he who follows the spirit of the warrior[110]
Hips, legs and fists, three unified into one
His spirit moves fast, his eyes are certain
The Kanji Gi in his name means if you are not righteous in principle, then no matter how much you refine your skill, it will all be in vain.
The Kanji Chin in his name means to not chase after what is rare or new,
For nothing can be accomplished unless you pursue it over a long time.

-Uemura Tsunekore
(Father of Uemura Wado?)
In celebration of Funakoshi Sensei's sixtieth birthday
1933

110 The author is referring to the fact that the Kanji 武 is composed of two elements 止 (tome/shi - "to stop") and 戈 (hoko - "spear/halberd"). The phrase is often interpreted as "stopping the spear2 or "using the spear to stop the spear," implying the purpose of martial arts is to prevent conflict, promote peace or stop aggression.

In celebration of Mr. Funakoshi Gichin Turning Sixty Years Old
By Rear Admiral Kanna Yorin

As Mr. Funakoshi is celebrating Kanreki,[111] a small publication related to Karate is being published so I was asked to contribute. Since Mr. Funakoshi looks to be no more than fifty years old, it is a total shock that he is approaching Kanreki. At the same time When I consider how he attained such health, I cannot help but nod and say, "So Karate is the reason."

Over the past fifty years, Mr. Funakoshi has devoted himself to discovering and training the inner mysteries of Ryukyu martial arts, succeeding in expanding the art so it is being practiced all over the country. Recently in the eleventh year of Taisho, he transmitted his knowledge in this art to students and youth in the Imperial Capital. Thanks to his efforts, Karate has been introduced to both the martial arts and sports world. Thus we have come to declare that he should be honored as the man who revitalized Karate.

It is my belief that the reason a martial art is respected and valued is not for its techniques but for its spirit and lessons on mental training. I believe it was for this reason that from days long past those referred to as Shihan, have been revered by the common people. I am not in the least ashamed to state that Mr. Funakoshi's refined manner and character place him on an equal level with those masters of old. It is no wonder then, that in this age his disciples are so numerous and distinguished.

At a time when both a diplomatic crisis and an economic crisis confront the nation, when I reflect deeply on the young people upon who's shoulders the success or failure of our nation rest, I cannot help but shudder with apprehension.

Ah, how truly sincere and strong he is. We place ever greater hopes in Mr. Funakoshi's future. May he conduct himself with prudence and take good care of himself.

---

[111] *Kanreki* 還暦 Refers to a person's 60th birthday (or 61st in the traditional age reckoning system) when one has lived through a full sexagenary cycle.

日大屋上庭園に於ける唐手練習

Karate training on the rooftop garden of Nichidai University

Made in the USA
Monee, IL
04 November 2026

67377473R00125